VIROLOGY MONOGRAPHS

DIE VIRUSFORSCHUNG IN EINZELDARSTELLUNGEN

CONTINUING / FORTFÜHRUNG VON
HANDBOOK OF VIRUS RESEARCH
HANDBUCH DER VIRUSFORSCHUNG
FOUNDED BY / BEGRÜNDET VON

R. DOERR

EDITED BY / HERAUSGEGEBEN VON

S. GARD · C. HALLAUER · K. F. MEYER

2

1968
Springer Science+Business Media, LLC

THE SIMIAN VIRUSES

BY

R. N. HULL

RHINOVIRUSES

BY

D. A. J. TYRRELL

1968

Springer Science+Business Media, LLC

ISBN 978-3-662-38598-2   ISBN 978-3-662-39447-2 (eBook)
DOI 10.1007/978-3-662-39447-2

© 1968 Springer Science+Business Media New York
Originally published by Springer-Verlag New York in 1968.
Softcover reprint of the hardcover 1st edition 1968
Library of Congress Catalog Card Number 68-26921

Printer: Steyrermühl, A-1061 Wien, Austria
Title No. 8329

The Simian Viruses

By

Robert N. Hull

Lilly Research Laboratories, Indianapolis
Indiana, U.S.A.

With 4 Figures

Table of Contents

I. Introduction

The extensive use of the techniques of tissue culture during the past 15 years has permitted the isolation of many new viral agents both from man and lower animals. These viruses have been isolated either by the inoculation of body fluids, excreta, or tissue homogenates, into previously prepared cultures, or by direct isolation of viruses from cultures prepared from apparently healthy animal tissues. This latter technique has been referred to as "unmasking of viruses". It was first demonstrated by the recovery of adenoviruses from cultures of "normal" human adenoid tissue (111), but we now know that a number of viruses can remain latent in tissues, and are undetectable by ordinary methods of virus isolation. When cells from such tissues are cultured *in vitro*, these latent viruses are freed, or released, from whatever bond (specific antibody, interferon) has held them in a state of equilibrium with their host. The virus so "freed" may produce a cytopathic effect (CPE), which indicates its presence, but, in some instances, such as in the case of rubella or SV_5 viruses, it may be necessary to use other technics to detect the presence of the virus. By these methods, virologists have recovered many new agents from both healthy and sick individuals, but in the latter case, it frequently was difficult to determine whether or not the virus isolated was the cause of the illness. Such determination was further complicated by the fact that the virus which was thought to have been isolated from the patient, may have been a latent virus in the tissue culture cells employed in the isolation studies. The virologists, therefore, have experienced essentially the same dilemma as the early bacteriologists, who attempted to determine the etiology of infectious diseases before they were fully aware of the "normal flora" of the species under study. Although the bacteriologist picked up contaminants on his lifeless media, and these resulted in misinterpretations, he did not have quite the same problem that confronts the virologists in relation to the presence of latent viruses in his culture systems.

Throughout the 1950s and 1960s, many virologists contributed toward the unraveling, systematic study and classification of the many new groups of viruses isolated from man and the lower species. Of the viruses isolated from man, we now recognize large groups of agents which can be placed in common families, such as adenoviruses, enteroviruses (with several sub-divisions) reoviruses, myxoviruses, herpesviruses, poxviruses, et cetera. Although there is yet much to be learned, it already has become apparent that groups of similar viruses are indigenous in sub-human primates and even lower animals (70). It has come to light, for example, that many species harbor viruses belonging to the adenovirus family. These viruses all possess the same general chemical and physical characteristics, including a common CF (complement fixing) group antigen. They are, however, serologically type specific, and in general are only infectious for their native host.

A similar parallel can be drawn with other major groups of viruses. This finding, again, is much like those made previously in the field of bacteriology, where it was demonstrated that most animals carried similar types of organisms in their intestinal tracts, or other tissues, but that antigenic differences exist between organisms isolated from different species. The streptococci offer a good example, in that Group A streptococci are common to man, while animals harbor the serological Group C streptococci. With the increased knowledge of the "normal viral flora" of man, and of the animal species which he uses in his studies, the virologist is now more competent to interpret, or to determine, the etiological significance of agents which he recovers from clinical specimens.

Monkeys and cultures of monkey cells have been used more widely in virological studies in recent years than any other animal species. Two Asiatic species, the rhesus and cynomolgus monkeys, were employed extensively in the 1950s, principally for research, production, and testing of poliomyelitis, adenovirus, and measles virus vaccines. In these efforts, thousands of monkeys and monkey kidney tissue cultures were used, and in the course of this work, many new viral agents were encountered. Although the complete story cannot be written, this chapter endeavors to bring together what is known about the "simian viruses", and to provide some information relating to their recognition, their significance in the monkey colony, and to the laboratory scientist.

Table 1. *Original Cytopathologic Classification of Simian Viruses*

CPE Group I	SV_1, SV_{11}, SV_{15}, SV_{17}, SV_{20}, SV_{23}, SV_{25}, SV_{27}, SV_{30}, SV_{31}, SV_{32}, SV_{33}, SV_{34}, SV_{36}, SV_{37}
CPE Group II	SV_2, SV_{16}, SV_{18}, SV_{19}
CPE Group III	SV_4, SV_{12}, SV_{28}, SV_{59}
CPE Group IV Miscellaneous	SV_5, SV_6, SV_{26}, SV_{29}, SV_{35}

Many of these agents were isolated, or studied and classified by HULL et al. (54, 55, 56) but many other investigators have contributed to the present knowledge. SABIN's (117) isolation and characterization of B virus in 1934, was the first report of the isolation of a virus indigenous in rhesus monkeys. The "foamy viruses" commonly found in uninoculated cultures of rhesus monkey kidney cells, were described by RUSTIGAN et al. in 1955 (116). As many additional agents were isolated in the early phases of the poliomyelitis virus vaccine program, an attempt was made to place them into groups in order to facilitate identification. The first obvious property of the viruses was the type of CPE which they produced in primary cultures of rhesus monkey kidney cells. Thus, in the first report by HULL et al. (54) the 10 agents described at that time were placed into 4 CPE groups. In a later publication, these groups were expanded to include additional viruses (56). This original classification by CPE groups is seen in Table 1. As additional agents were encountered, many fitted into these same groups, but others required the establishment of new CPE groups, and as more information became available, some of those agents listed in the miscellaneous category, were moved to the other more specific groups. When criteria were established to classify viruses into families such as adenoviruses, enteroviruses, et cetera, it was obvious that the various CPE groups of simian viruses possessed characteristics which allowed them to be classified into such recognized groups of viruses. Thus, all agents included in CPE Group I, for example, were found to be adenoviruses. With the greater knowledge now available, the simian viruses should be reclassified, or regrouped, as suggested in Table 2 according to the general scheme suggested by ANDREWES et al. (1). This classification with nine groups of viral agents accommodates not only the agents with SV designations, but also permits

Table 2. *Proposed New Classification of Simian Viruses*

DNA Ether Resistant Viruses

Adenoviruses	SV_1, SV_{11}, SV_{15}, SV_{17}, SV_{20}, SV_{23}, SV_{25}, SV_{27}, SV_{30}, SV_{31}, SV_{32}, SV_{33}, SV_{34}, SV_{36}, SV_{37}, SV_{38}, SA_7
Papovaviruses	SV_{40} (SA_{12})[1]

DNA Ether Sensitive Viruses

Herpesviruses	
Group A	B-Virus, SA_8, MHV, SMV
Group B	Cytomegaloviruses SA_6
Poxviruses[2]	YABA, Monkey pox, YLD

RNA Single Stranded Ether Resistant Viruses

Picornaviruses	
Enteroviruses	SV_2, SV_6, SV_{16}, SV_{18}, SV_{19}, SV_{42}, SV_{43}, SV_{44}, SV_{45}, SV_{46}, SV_{47}, SV_{48}, SV_{49}, SA_5
Enteroviruses?	SV_{26}, SV_{35}
Not Enteroviruses	SV_4, SV_{28}, SA_4

RNA Single Stranded Ether Sensitive Viruses

Myxoviruses	
Group A	Hemadsorption positive SV_5, SV_{41}
Group B	Hemadsorption negative FV_1, FV_2, FV_3, FV_4, FV_5, SA_1

RNA Double Stranded Ether Resistant Viruses

Reoviruses	SV_{12}, SV_{59}, SA_3

[1] Questionable classification.
[2] Ether sensitivity is questionable.

the inclusion of other known viruses of simian origin, such as B virus, the "foamy viruses", and pox viruses. In the SV series of viruses, the missing numbers were held by isolates which were later found to be identical with previously isolated viruses, or were found not to be viruses at all. SV_{24}, for example, proved to be an amoeba of the genus *Acanthamoeba*.

II. Simian Adenoviruses

The viruses listed in CPE Group I produced a rounding of cells and formation of grape-like clusters when inoculated into cultures of monkey kidney cells (Fig. 1), which very closely resembled the CPE produced by human strains of adenoviruses in the same culture system. The viruses were not neutralized by antiserum prepared against the more common strains of human adenoviruses, but it was found that the simian variety contained the CF group antigen of the human strains. Electron micrographic studies have also revealed that the simian adenoviruses are 70 to 80 millimicrons in diameter, have a polyhedral shape, and an arrangement of subunits, with a 5:3:2 symmetry. The subunits are 50 to 60 angstroms in diameter. Such observations were reported, first, by TYRRELL et al. (132), in studies with SV_{17}, but were described in greater detail by ARCHETTI et al. (2), who studied a virus identified at that time as SV_{39}. In a second report, ARCHETTI and STEVE-BOCCIARELLI (3), extended these observations to include SV_1, SV_{11}, SV_{20}, SV_{23}, and SV_{34}. At this time, they concluded that the virions were regular icosahedrons with a 5:3:2 symmetry, and contained a total of 252 capsomeres. These observations agreed with those of HORNE et al. (49), for human adenovirus type 5. REIMER (107) studied all of the SV classified simian adenoviruses by electron-micrographic methods, and confirmed that they have the dimensions and morphology as described by ARCHETTI. Representative electronmicrographs are seen in Fig. 2.

In experiments not reported in the literature, the author and his associates found that all of the simian adenoviruses possessed DNA in their nucleic acid core. These determinations were made by inhibition tests with a DNA inhibitor, FUDR. The viruses were also shown to be resistant to ether. Thus, the CPE Group I simian viruses met the physical and chemical criteria for adenovirus classification.

Simian adenoviruses also share many biological properties in common with the human strains. They are commonly isolated from the respiratory and intestinal tracts of monkeys, but like the human strains, certain types are more prevalent in the respiratory tract than others. In general, SV_1, SV_{11}, SV_{15}, SV_{17}, SV_{23}, and SV_{32}, would be more likely to occur in the respiratory tract than the other members of the group, many of which were stool isolates. TYRRELL et al. (132) reported the isolation of SV_{17} from patas monkeys in a colony experiencing an epidemic of conjunctivitis and rhinorrhoea. In an investigation of conjunctivitis in rhesus monkeys, the author and associates found that about one-half of the non-bacterial infections were due to simian adenoviruses, and that the predominant agent isolated was SV_{32}. BULLOCK (14) reported further similarities between the human and simian adenoviruses in isolations made from the nose and throat of monkeys and by the unmasking of latent viruses from monkey tonsillar tissue. SV_{17} and SV_{32} were isolated from nose and throat swabs of monkeys with acute respiratory illness, and the same viruses, plus SV_{15}, were also recovered from healthy animals. The same three viruses were recovered from cultures of monkey tonsil tissue.

The experimental infection of vervet monkeys with SV_{17} and its similarity to the disease seen in man infected with adenoviruses was reported by HEATH

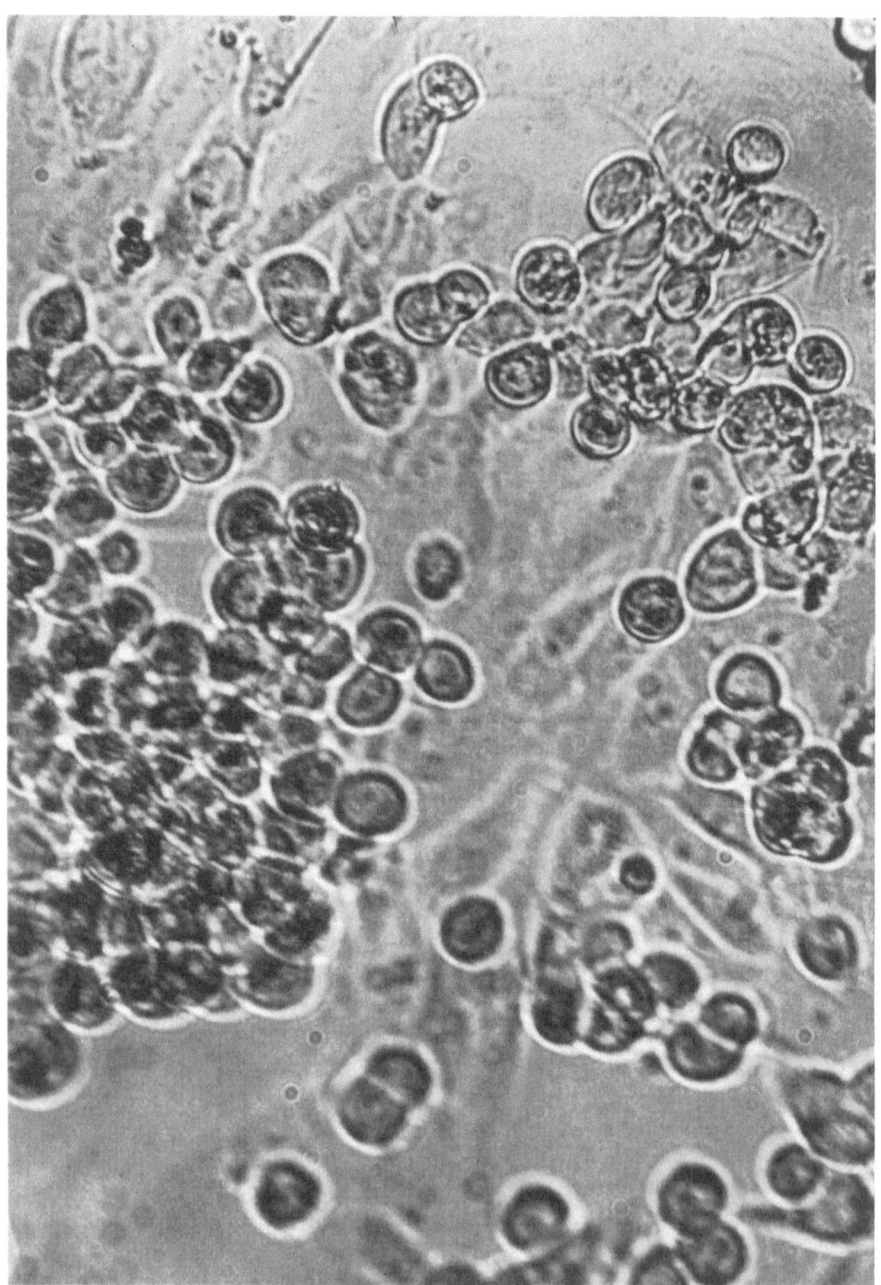

Fig. 1. Simian Adenovirus, SA₄, CPE in Vervet monkey kidney cells. 150 × before enlargement.

et al. (41). Six antibody negative (by both HI and neutralization tests) monkeys were given a dose of $10^{6.7}$ $TCID_{50}$ of virus by the intranasal route. All became infected as judged by clinical symptoms, virus recovery, and antibody response. The principle symptoms were reddening and swelling of the pharyngeal mucous membranes and the tonsils. This was similar to the findings in humans infected with adenoviruses. It was noted in the monkeys, also, that spleen enlargements followed infection with SV_{17} virus. Virus was excreted in the nose and throats from three to nine days after infection. Post infection antibody titer by HI test ranged from 1:64 to 1:256 and from 1:48 to 1:512 by virus neutralization tests.

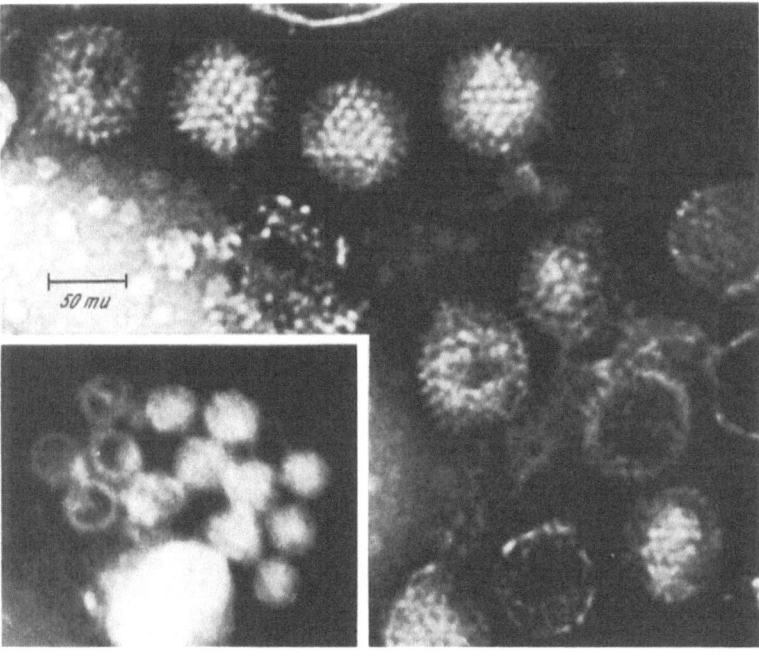

Fig. 2. Electronmicrographs of Simian Adenovirus, SV_{37}, and an AAV (adeno associated virus) isolated from Simian Adenovirus, SV_{15}. Negative stained with phosphotungstate. × 200,000.

There has been no evidence that vervet monkeys experience natural infection with the rhesus monkey adenovirus, SV_{17}, but these experimental infections point out their potential susceptibility to this agent.

HULL et al. (61) described the oncogenic properties of certain of these simian adenoviruses for newborn hamsters, which, again, paralleled findings with strains of human origin. Five of the 17 adenoviruses in the SV series were found to produce tumors following subcutaneous inoculation, as well as SA_7 isolated from the African green monkey. This group included SV_{20}, SV_{33}, SV_{34}, SV_{37}, and SV_{38}. None of the viruses commonly found in the respiratory tract of monkeys proved to be oncogenic. The tumors produced by the viruses were undifferentiated neoplasms with some characteristics of lymphomas of the reticulum-cell type, and in this respect, were like those produced by human adenoviruses. Histologically,

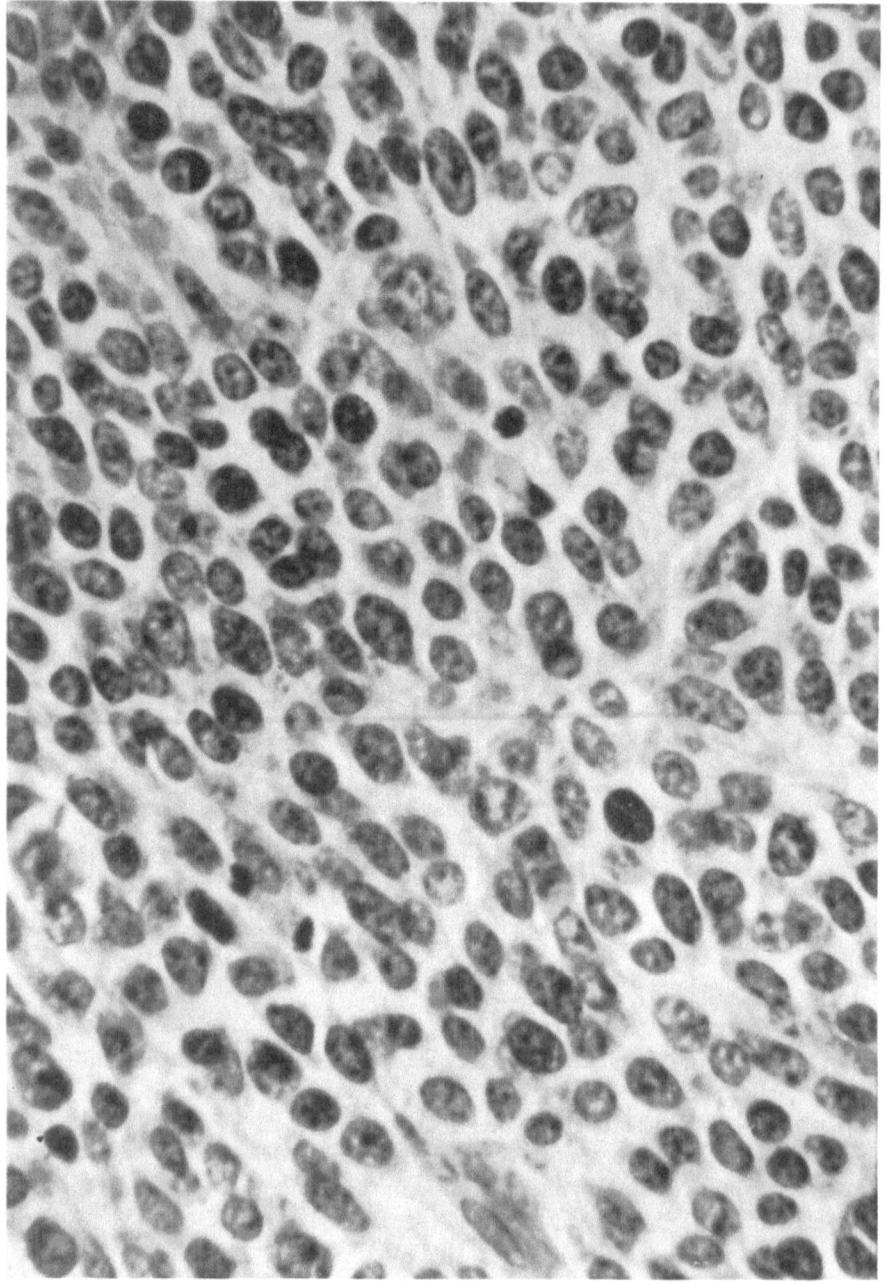

Fig. 3. Hematoxylin and eosin stained section of a lymphoma-like hamster tumor produced by Simian Adenovirus, SA₇. Approx. 400 ×.

the tumors were quite unlike the fibrosarcomas produced by SV_{40} virus (Fig. 3). The tumors contained a "T" antigen which reacted in CF tests only with sera from animals bearing tumors produced by the homologous viruses. There were no cross-reactions amongst the "T" antigens of the tumors produced by the different simian adenoviruses, nor did they react with those produced by human adenoviruses or by SV_{40}.

The simian adenoviruses share yet another property in common with the human types, in that some of them have a smaller associated virus or particle called AAV, Adeno Associated Virus, by ATCHINSON et al. (4). REIMER (107) studied all the adenoviruses in the SV series except SV_{36}, by electron micrographic methods and found that five of them, SV_1, SV_{11}, SV_{15}, SV_{25}, and SV_{34} contained a smaller particle in addition to the 70 mμ adenoviruses, and suggested that these could be viral in nature or sub-units of the adenovirus. These particles ranged in size from 15—20 mμ (those seen in preparations of SV_1) to 30 mμ (seen in electron micrographs of SV_{15}). A thorough search of many electron microscopic fields of preparations of all the other simian adenoviruses failed to detect these smaller particles seen in the above-mentioned strains. The report by ATCHINSON et al. (5) presented a very thorough and comprehensive study of the smaller particle associated with SV_{15} virus. These authors describe the virus as a hexagonal, 24 mμ particle, containing double stranded DNA. By centrifugation and filtration they were able to obtain purified preparations of AAV free of SV_{15} virions. Such material failed to proliferate, or to produce CPE in monkey kidney culture, and in many other cell types tested. When the pure AAV preparation was inoculated into cultures along with an AAV free strain of SV_{15}, multiplication of AAV occurred. Further, they demonstrated that AAV could be cultivated, using human adenovirus type 2 as the "helper virus". Many other DNA and RNA containing viruses capable of growth in monkey kidney cultures failed to serve as a "helper virus" for growth of AAV. Antiserum was prepared against the pure AAV preparation which inhibited AAV, but not SV_{15}. They also failed to show any antigenic relationship between AAV and SV_{15} by CF and precipitation tests. These authors concluded that AAV was an incomplete or defective virus.

A report by MAYOR et al. (86) appeared at about the same time as the above-described paper which also dealt with a small particle found in SV_{15} virus suspensions. These authors described it as 20 mμ polyhedral particle with a density of 1.43. These particles contained protein and double stranded DNA. They had a cubic symmetry of the icosahedral type with a coat composed of 12 sub-units, each at the vertex of an icosahedron. They found also that the particles had a hemagglutinin which was inhibited by SV_{15} antiserum, and that it fixed complement with a human adenovirus reference serum. These authors, therefore, concluded that it was an internal component of the mature SV_{15} virion, and was related to the adenovirus genome.

In a subsequent paper, ATCHINSON et al. (5) described further electron microscopic studies of the AAV agent and reported that replication of AAV and SV_{15} occurred simultaneously in the nucleus of a single cell. In such doubly infected cells, however, one or the other of the two viruses was predominant in number. These data were presented as further evidence of their contention that the AAV particle is a separate entity and not a sub-unit, or part of SV_{15} virus.

RAPOZA (106) has made some interesting observations relating to the inter-
actions, or relationships between AAV and the simian adenoviruses. He reports
that simian adenovirus seed pools should be free of AAV if they are to be employed
in hemagglutination tests, as AAV inhibits the production of adenovirus hem-
agglutinins. SV_{36} has been a difficult virus to propagate as its growth has been
limited to primary rhesus monkey kidney culture and virus yields have been
low, rarely exceeding titers of $10^{-3.0}/0.5$ ml of culture fluid. RAPOZA states, however, that SV_{36} is heavily contaminated with AAV particles, and once the culture
is freed of AAV, that it grows well in strain $LLC-MK_2$ (59) or in primary African
green monkey kidney cultures. Yields of SV_{36} under these conditions were as
high as $10^{-6.0}/0.1$ ml. (This has not yet been confirmed in the author's laboratory,
although RAPOZA indicated that it had been noted independently by CASTO at
the University of Pittsburgh.) Both RAPOZA and CASTO stated further that the
presence of AAV in the culture of an oncogenic adenovirus inhibited the ability
of the virus to produce tumors.

The primary concern of many investigators with simian adenoviruses (as well
as with other types of simian viruses) lies in their need to distinguish them from
adenoviruses of human origin. There is no certain way of doing this short of making a serological identification. The pattern of growth, however, in monkey
and human cells gives a degree of presumptive evidence. All of the simian adenoviruses will grow in rhesus monkey kidney cells in the absence of SV_{40} and all
but SV_{36} (data from author's laboratory) will grow in strain $LLC-MK_2$. Of the
more common types of human adenoviruses only type 2 will grow in strain LLC-
MK_2 cells, and as had been reported by BEARDMORE et al. (10) and others,
at least some types of human adenovirus cannot be cultivated in monkey kidney
cultures unless SV_{40} virus is present. Further, we have found (54, 56) that only
SV_1, SV_{11}, SV_{25}, SV_{33}, and SV_{34}, will grow in cells of human origin while the other
serotypes failed to grow. (Data to the contrary relating to SV_{36} found in Reference 56 has since been disproved.) Thus, by determining the pattern of growth
in the cell systems mentioned, it is possible to make at least a tentative determination of the type of adenovirus with which one is dealing.

Serological identification is best accomplished by quantitative serum neutralization tests, and for some types, cross-neutralization tests are necessary. Hemagglutination-inhibition tests may also be employed. A number of cross-reactions
are seen within the simian adenovirus group, and while some of these are fairly
consistent from one set of typing sera, to another, a great deal of variation is
also encountered, both in the degree of cross-reaction and in respect to which
agents demonstrate such reactions. The practice of giving multiple injections
of antigens and/or of holding immunized animals as a source of typing sera and
periodically giving booster doses tends to broaden the antigenic spectra of the
viruses and this results in more extensive cross-reactions. More type specific
sera are obtained when reduced amounts of antigen are administered; however,
as might be expected, the titers of these sera are lower. Cross-reaction data obtained with such sera are seen in Table 3. These data, as well as similar findings with
numerous sets of typing sera indicate two groups of viruses which consistently
show some degree of cross-reactions within the groups. One such group (Group A)
includes SV_{15}, SV_{17}, SV_{27}, and SV_{31}; the other group (Group B) involves SV_1,

SV_{33}, SV_{34}, and SV_{38}. Earlier findings reported in Reference No. 56 are reproduced in Table 4 for comparison with the more recent cross-reaction data seen in Table 3. SV_{15} and SV_{17} generally have shown cross-reactions, while SV_{27} and SV_{31} are either identical, or very closely related. The data in Table 3 suggests that they are identical, while the earlier results seen in Table 4 indicate a very close serological relationship, with the only difference being that SV_{31} was neutralized by anti-SV_{27} sera to a 16-fold lower titer than was the homologous virus. SV_{31} antiserum neutralized both viruses to the same terminal dilution. Although some cross-reactions have been noted between the SV_{15} and SV_{17} pair of viruses, and SV_{27} and SV_{31}, these reactions have been far less significant than those noted between the two members of each pair of viruses. Although not documented in Tables 3 and 4, the results of many cross-neutralization tests over the years have suggested that SV_{15} and SV_{17} were more closely related to SV_{27} than to SV_{31}.

SV_{1} and SV_{34}, as seen in both Tables 3 and 4, showed some degree of cross-reaction which was intensified in high titered sera such as that listed in

Table 3. Cross-neutralization Tests with Simian Adenoviruses Sera

Viruses	SV_{1}	SV_{11}	SV_{15}	SV_{17}	SV_{20}	SV_{23}	SV_{25}	SV_{27}	SV_{30}	SV_{31}	SV_{32}	SV_{33}	SV_{34}	SV_{36}	SV_{37}	SV_{38}
SV_{1}	64	32[1]	—	—	—	—	—	—	—	—	—	—	16	—	—	—
SV_{11}	—	256	—	—	—	—	—	—	—	—	—	—	—	—	—	—
SV_{15}	—	—	256	64	—	—	—	—	—	—	—	—	—	—	—	—
SV_{17}	—	—	—	256	—	—	—	—	—	—	—	—	—	—	—	—
SV_{20}	—	—	—	—	128	—	—	—	—	—	—	—	—	—	—	—
SV_{23}	—	—	—	—	—	32	—	—	—	—	—	—	—	—	—	—
SV_{25}	—	—	—	—	—	—	256	—	—	—	—	—	—	—	—	—
SV_{27}	—	—	32	—	—	—	—	256	—	64	—	—	—	—	—	—
SV_{30}	—	—	—	—	—	—	—	—	512	—	—	—	—	—	—	—
SV_{31}	—	—	—	—	—	—	—	64	—	256	—	—	—	—	—	—
SV_{32}	—	—	—	—	—	—	—	—	—	—	256	—	—	—	—	—
SV_{33}	—	—	—	—	—	—	—	—	—	—	—	256	—	—	—	—
SV_{34}	—	—	—	—	—	—	—	—	—	—	—	—	128	—	—	64
SV_{36}	—	—	—	—	—	—	—	—	—	—	—	—	—	1024	—	—
SV_{37}	—	—	—	—	—	—	—	—	—	—	32	—	—	—	256	—
SV_{38}	—	—	—	—	—	—	—	—	—	—	—	32	—	—	—	256

Note — = <1:16.
[1] This neutralization of SV_{1} by anti SV_{11} has not been a consistent observation.

Table 4. The lower titered sera used in Table 3 (which, as mentioned above, was prepared by immunization with fewer doses of antigen) failed to show these higher levels of cross-reaction. SV_{33} and SV_{34} revealed antigenic similarity in both sets of typing sera, but only SV_{34} showed any serological relationship to SV_1. SV_{38} had not yet been isolated when the data presented in Table 4 were recorded. In Table 3, it is seen that a strong, but distinguishable, serological relationship exists between SV_{33} and SV_{38}. SV_{38} appeared to be unrelated to SV_1 and SV_{34}.

Table 4. *Cross Reactions of Selected Simian Adenoviruses*[1]

Viruses	Sera				Viruses	Sera		
	SV_{15}	SV_{17}	SV_{27}	SV_{31}		SV_1	SV_{33}	SV_{34}
SV_{15}	2048	4	<4	16	SV_1	4096	4	64
SV_{17}	16	256	16	16	SV_{33}	<4	1024	64
SV_{27}	256	64	1024	1024	SV_{34}	512	16	128
SV_{31}	<10	<10	64	1024				

[1] Extracted from an earlier publication (Ref. No. 56, Table 4)

Table 5. *Identification of Simian Adenoviruses by Intersecting Serum Pools*

Pools	Sera contained in pools
A	SV_1, SV_{11}, SV_{15}, SV_{17}, SV_{20}, SV_{23}
B	SV_1, SV_{25}, SV_{27}, SV_{30}, SV_{31}, SV_{32}
C	SV_{11}, SV_{25}, SV_{33}, SV_{34}, SV_{36}, SV_{37}
D	SV_{15}, SV_{27}, SV_{33}, SV_{38}
E	SV_{17}, SV_{30}, SV_{34}, SV_{38}
F	SV_{20}, SV_{31}, SV_{36}
G	SV_{23}, SV_{32}, SV_{37}

Neutralized by pools	Key to identification	Neutralized by pools	Key to identification
	Virus identification		Virus identification
A+B	SV_1	B+E	SV_{30}
A+C	SV_{11}	B+F	SV_{31}
A+D	SV_{15}	B+G	SV_{32}
A+E	SV_{17}	C+D	SV_{33}
A+F	SV_{20}	C+E	SV_{34}
A+G	SV_{23}	C+F	SV_{36}
B+C	SV_{25}	C+G	SV_{37}
B+D	SV_{27}	D+E	SV_{38}

The new classification of the simian viruses seen in Table 2 did not include SV_{39} in the simian adenovirus group. This virus, isolated in our laboratory, has never been described in the literature by the author, but it has been mentioned by others (2) and has been discussed at several informal meetings on simian viruses. Since its original identification and assignment of an SV number, it has been found to be identical with SV_{23}, and, therefore, has been dropped from the simian adenovirus classification.

In spite of the cross-neutralization amongst some members of the simian adenovirus group, a scheme has been devised for more rapid identification by the use of intersecting serum pools. This method, as outlined in Table 5, was developed in the author's laboratory and has proved to be valuable in quickly determining the likely identity of a new isolate, or unknown virus. When coded samples of virus were supplied to two different laboratories as unknowns, both were able to identify the virus in each of the samples using these pooled sera. As a matter of practice, however, a virus tentatively identified by this method is assayed further with specific monovalent serum in order to confirm the identification.

RAPOZA and CHEEVER (105) have studied the hemagglutination (HA) patterns of the simian adenoviruses and have performed hemagglutination-inhibition (HI) tests amongst the member viruses. These authors have proposed three HA groups based on the ability of the virus to hemagglutinate rat, guinea pig, or rhesus monkey red blood cells (RBC) at either 4°C or 37°C. These HA groups are seen in Table 6. It was interesting to note that all of the oncogenic types of

Table 6. *Simian Adenovirus Hemagglutination Groups* (RAPOZA *and* CHEEVER)

Rhesus RBC only at 4°C and 37°C	Rat[1], rhesus, G. pig RBC at 4°C	Incomplete agglutination of rat RBC at 37°C[2]	No agglutination of rat, rhesus or G. pig RBC
Group I	Group II	Group III	Group IV
SV_{36}	SV_{15} SV_{17} SV_{23} SV_{27} SV_{31} SV_{32} SV_{37} SV_{39}	SV_1 SV_{11} SV_{20} SV_{25} SV_{30} SV_{33} SV_{34} SV_{38}	SA_7

[1] Agglutination of rat cells is incomplete and not temperature dependent.

[2] These viruses also hemagglutinate rhesus and guinea pig cells at 4°C, except for SV_{11}, SV_{25}, and SV_{30}. Agglutination of rat cells at 4°C is incomplete for all viruses except for SV_{25} and SV_{30}.

simian adenoviruses fell into HA Group III except for SV_{37}, and that most of those generally associated with respiratory diseases in the monkey were included in HA Group II. No cross-reactions were obtained between viruses listed in separate groups, but some reactions were noted amongst the viruses within Group II or Group III. In Group II they found that SV_{27} and SV_{31} were identical, which agrees with the neutralization test data in Table 3. They also found that SV_{23} and SV_{39} were very closely related, but felt that they could distinguish between them. This finding is in agreement with our earlier experience with these two viruses, although more recent data obtained from neutralization tests indicated they were identical. Thus, the elimination of SV_{39} from the classified listed simian adenoviruses may be premature. By HI tests, RAPOZA and CHEEVER found SV_{32} and SV_{37} to be identical, but stated that this was not true in virus neutralization tests. They found, also, minor HI cross-reactions amongst the SV_{15}, SV_{17}, SV_{27}, and SV_{31} group of viruses much as previously noted in neutralization tests with these viruses. In the HA Group III viruses, they reported that SV_{33} and SV_{38} were identical, both by HI and serum neutralization tests. Data in Table 3, however, indicated these viruses to be very closely related, but not identical. A minor one way cross-reaction was seen between SV_{34} and SV_{38}, and this, too, is in keeping with the data presented in Table 3. With few exceptions, therefore, the serological interrelationships amongst the simian adenoviruses were found to be the same by either serum neutralization or HI tests.

Although neutralizing antibodies against some of the simian adenoviruses have been found in human sera, or human γ-globulin, there are no significant cross-reactions between the simian and human strains of adenoviruses either by serum neutralization or HI tests. These studies were undertaken in a joint effort by the author and Dr. J. L. SEVER (N.I.H.). All of the simian adenoviruses and their respective antisera were included in these studies, and of the 28 types of human adenoviruses available at that time, all, except for types 19 through 24 were included. Reference sera prepared against the human strain were used and included all types but 8, 14, 18, and 20. Neutralization tests were performed in the author's laboratory with these reagents, while Dr. SEVER carried out the HI tests. In the neutralization tests, SV_{11} antiserum neutralized adenovirus type 25 at 1:10, but not at a 1:20 dilution, and the same level of neutralization was noted with SV_{25} serum against adenovirus type 12. None of the antisera prepared against the human strains neutralized any of the simian adenoviruses. Equally negative results were obtained by the HI test method.

The simian adenoviruses are stable at $-70°C$ for years, for many months at $4°C$, and will even survive for several weeks at $37°C$. They can be preserved by lyophilization and are stable in this state either at $4°C$, or room temperature. It is possible, therefore, to ship them over long distances without refrigeration if some drop in titer can be accepted.

III. Simian Picornaviruses

The simian Picornaviruses have received far less attention in the literature than have the adenoviruses. This, perhaps, is because they rarely occur as latent viruses in monkey kidney cultures (except for SV_4 and SV_{28}), and when they are isolated from such cultures, it most often is the result of fecal contamination at the time of kidney removal. Although they are readily isolated from monkey stool specimens (except for SV_4 and SV_{28}), they are rarely recovered from other tissues, and, again, when such isolation is made, one should consider fecal contamination. At least in respect to SV_2, few monkey sera are found to contain antibody which further suggests that these viruses do not invade tissue other than the intestinal tract. In unpublished studies, vanFRANK (133) recovered numerous strains of simian enteroviruses by exposing Petri dish cultures of monkey kidney cells to the air in large monkey storage rooms. No other types of simian viruses were isolated.

The CPE of the first four viruses isolated, SV_2, SV_{16}, SV_{18}, and SV_{19} suggested that these viruses might belong to the enterovirus family. SV_2, SV_{16}, and SV_{18} were originally recovered from cultures of monkey kidney cells, but SV_{19} was a stool isolate recovered by HOFFERT, BATES and CHEEVER (44). SV_6, originally was not included in the same group with these four viruses, since its CPE and growth characteristics were considerably different. It was known, however, that it was a frequent virus in monkey stool samples, and, therefore, in the new classification (Table 2) it has been placed in the enterovirus group, along with the other agents mentioned above. SV_{29} described in previous reports (56) has since been shown to be identical to SV_6, and has, therefore, been removed from the classification. SV_{26} was originally recovered from the central nervous system (CNS) tissues of a rhesus monkey, but its size and other properties suggested that it, too, belonged in the Picornavirus group. SV_{35} also was isolated from monkey CNS tissues, but further studies of its properties indicated that it was a member of the Picornavirus group. Neither SV_{26} nor SV_{35} have been isolated from stool samples. The newest members of this group, SV_{42} through SV_{49}, were isolated and characterized by HEBERLING and CHEEVER (42), and have not been studied in the author's laboratory. These authors have made the most thorough study of the simian enteroviruses published to date. Their findings will be discussed subsequently.

In addition to the cytopathic effect which these viruses produce in tissue culture and their frequent isolation from monkey stool specimens, their physical and chemical properties are like those of the members of the Picornavirus family.

Table 7. *CPE and Plaque Formation Grouping of Simian Enteroviruses* (HEBERLING *and* CHEEVER)

Groups	Viruses
A	SV_6, SV_{19}, SV_{43}, SV_{44}, SV_{46}, SV_{47}, SV_{48}
B	SV_{16}, SV_{18}, SV_{42}, SV_{45}
C	SV_2
D	SV_{49}

They are ether resistant, and contain RNA in their nucleic acid core as determined by specific nucleic acid inhibitors. By ultrafiltration experiments, ATOYNATAN and HSIUNG (6) found SV_{16}, SV_{18}, SV_{26}, SV_{29} (identical to SV_6), and SV_{35}, to be smaller than 50 mμ, which placed them in the general size range of the human enteroviruses. HEBERLING and CHEEVER (42) found by electron microscopic studies that SV_2, SV_{18}, SV_{19}, SV_{42}, SV_{45}, SV_{48}, and SV_{49} were uniform in size between 32.1 and 38.4 mμ. The viruses are resistant to low pH, which distinguishes them from the rhinovirus group. SV_{19} and SV_{44} are virulent for suckling mice (44). SV_2 agglutinates rhesus RBC's at 4°C. HEBERLING and CHEEVER (42) confirmed this, and reported further that SV_{45} agglutinated African green monkey red cells at 4°C and SV_{46} agglutinated chicken red cells at 4°C. These simian enteroviruses in general more nearly resemble the Coxsackie B viruses, and some of the ECHO viruses, than they do polioviruses. Only their physical and chemical properties are similar to poliovirus as is true of all the members of the Picornavirus group.

The simian enteroviruses are not a homogenous group of agents as is readily demonstrated in the report by HEBERLING and CHEEVER (42). Based on growth properties in rhesus monkey kidney cultures alone, these authors have subdivided the group into four categories. This grouping is seen in Table 7. The members of these groups, however, did not share other properties in common such as

hemagglutination, ability to grow in New World, or African monkey kidney cultures or animal virulence. These workers also reported further chemical and physical properties of these viruses. They demonstrated stability in the presence of
molar magnesium chloride at a temperature of 50°C for one hour which further
characterized the viruses as enteroviruses. The viruses were resistant to photo-
inactivation when grown in the presence of toluidine blue, but were sensitive in
the presence of neutral red. Attempts to inhibit the simian enteroviruses by treatment with guanidine and HBB produced results which varied from those previously reported by TAMM and EGGERS (128) for strains of human enteroviruses.
Only the Group A viruses (Table 7) showed some inhibition by guanidine, while
HBB failed to inhibit any of the viruses. This latter finding was in keeping
with the properties of Group A Coxsackieviruses, but what resemblance these
simian enteroviruses have to the Coxsackieviruses, in all other respects reflects
Group B, rather than Group A.

Table 8. *Cross-neutralization Data Obtained with the First 7 Simian Picornaviruses*

Viruses	Antisera						
	SV_2	SV_6	SV_{16}	SV_{18}	SV_{19}	SV_{26}	SV_{35}
SV_2	128	$-^1$	—	—	—	—	—
SV_6	—	128	—	—	—	—	—
SV_{16}	—	—	128	—	—	—	—
SV_{18}	—	—	—	64	—	—	—
SV_{19}	—	—	—	—	32	—	—
SV_{26}	--	—	—	—	—	32	—
SV_{35}	—	—	—	—	—	32	226

$^1 = <1{:}16$.

Serological identification of some of the simian enteroviruses has frequently
caused difficulty due to cross-reactions, principally amongst the SV_2, SV_{16},
SV_{18}, and SV_{19} group. As with the simian adenoviruses, however, these cross-
reactions can be reduced or eliminated by immunization with lesser amounts of
antigen. Results obtained with such sera are presented in Table 8, where it is
seen that no cross-reactions occurred at the levels of antisera tested except for
SV_{26} and SV_{35}. Table 5 in reference (56) showed a greater degree of cross-reaction
when higher titered sera were employed. SV_2 appeared to be the broadest antigen
as its serum most frequently neutralized, to some extent, other viruses within
the group. SV_{16} and SV_{18} also cross-react when high titered antisera are used.
SV_{19} is the poorest antigen in the group; thus, its generally low titered antiserum
seldom neutralized the other viruses. SV_{19} virus, however, has been neutralized
at low levels by SV_2, SV_{16}, and SV_{18} antisera when the homologous titers of
those sera were high.

HEBERLING and CHEEVER (42) in their report did not give cross-reaction
data on all the viruses in the enterovirus group, but did present such data for
agents contained within their proposed CPE groups. Of the Group A viruses
(see Table 7) the most extensive cross-reactions occurred between SV_{19} and SV_{48}.
SV_{48} antiserum neutralized SV_{19} to within 2-fold of the homologous titer, although

SV_{19} antiserum failed to neutralize SV_{48}. Other cross-reactions amongst members of this group were of a minor nature. The viruses in Groups B, C, and D were examined collectively in serological studies. Data taken from a table in the manuscript by these authors (HEBERLING and CHEEVER) have been rearranged and are presented in Table 9. SV designations have been used for all the agents listed in the table to avoid confusion, and agents P11 and P17 included in the original manuscript were eliminated from consideration at the time, due to indecision over their identity. It is readily seen in Table 9 that SV_2 antiserum had the broadest antibody coverage as it neutralized SV_{16}, SV_{18}, and SV_{42} viruses, as well as the homologous SV_2 virus. SV_2 and SV_{16} showed two-way cross-reactions, as did also, SV_{42} and SV_{45}. SV_{45} and SV_{49} crossed one-way, as SV_{49} antiserum neutralized SV_{45} virus.

It is unfortunate that, as yet, all 15 of the simian enteroviruses have not been studied for cross-reactions at the same time in one laboratory. HEBERLING and CHEEVER may have done this, however, as they state in their publication that they observed no cross-reactions between members of their CPE Group A viruses and agents belonging to other groups.

Table 9. *Cross-neutralization Data Obtained from a Selected Group of Simian Enteroviruses* (HEBERLING and CHEEVER)

Viruses	Antisera					
	SV_2	SV_{16}	SV_{18}	SV_{42}	SV_{45}	SV_{49}
SV_2	40	$-^1$	—	—	—	—
SV_{16}	20	160	—	—	—	10
SV_{18}	160	40	80	10	80	—
SV_{42}	10	—	—	1280	80	—
SV_{45}	—	—	—	—	320	160
SV_{49}	—	—	—	—	—	1280

$^1 = <1:10.$

The simian enteroviruses are quite stable, surviving for years at $-70°C$ and for long periods at $4°C$. Lyophilization of SV_2, SV_6, SV_{16}, SV_{18}, SV_{19}, and SV_{26} has not been accomplished.

SV_4, SV_{28}, and SA_4 are listed under the Picornavirus group because SV_4 possesses the physical and chemical properties of this family of viruses. SV_{28} and SA_4 have not been studied as extensively as has SV_4, but because of the similar CPE which they produce in monkey kidney cultures and some serological relationships, it is assumed for the time being, that they belong in the same group as does SV_4. It is known that both SV_{28} and SA_4 are RNA, ether resistant viruses. Although both SV_4 and SV_{28} were frequently isolated from rhesus monkey kidney cultures, they were never isolated from stool samples, and it is for this reason that they are not included in the enterovirus group.

SV_4 was most thoroughly studied by TAYLOR (129) and by SATTAR and ROZEE (118). TAYLOR found by electron microscopic examination that SV_4 was a particle of $25-30$ mμ in diameter. It had a density of 1.31. The CPE, as evaluated in H and E stained thin sections examined by electron microscopy, consisted of margination of chromatin and vacuolization of the cytoplasm. These two groups of workers likened SV_4 to the enteroviruses, or the Columbia SK group, based on the above observation, plus stabilization by molar magnesium chloride and resistance to a pH of 3.5. They found the virus to be non-pathogenic for suckling mice and to agglutinate rhesus RBC at 4°C, but not human type 0, or sheep cells. Cross-neutralization tests with EMC and Mengo viruses of the Columbia SK group revealed no serological relationship of SV_4 to these viruses. SATTAR and ROZEE, based on their observation, concluded that SV_4 was not like the Columbia SK or Coxsackie groups of viruses, and suggested that it should be classified as an enteric cytopathogenic orphan virus of simian origin. While it is true that it has many properties in common with the human ECHO viruses, there remains the fact that apparently it is *not* an enteric virus, since it has never been isolated from the intestinal tract.

ROSEN (110) has suggested that SV_4 and SV_{28} may be reoviruses, but admits that they are distinguishable from the three established serological types of reoviruses. Other properties such as size, and lack of recovery from the intestinal or respiratory tracts would tend to contradict this. Further, these viruses do not produce the large globular cytoplasmic inclusion seen in reovirus infected cells. These viruses also probably contain single stranded RNA based on their rapid inactivation by formaldehyde.

During the mid-1950s, SV_4 was the most frequent contaminant occurring in rhesus monkey kidney cultures in our laboratory (SV_{40} was not yet known at this time, nor was hemadsorption used to detect SV_5). Many viruses received from other laboratories during this same period were found also to be SV_4. Of 918 simian virus isolations made between April 1, 1955, and October 1, 1957, 504 were identified as SV_4. Sixty-five of these were SV_{28}. During a later period of time, between January 1, 1958, and May 1, 1962 (when fewer rhesus monkeys were being used) 68 simian virus recoveries were made from uninoculated cultures. Of these, 15 proved to be SV_{28}, and only one was identified as SV_4. This experience has continued to the present time. SV_4, once the most commonly isolated virus, has practically dropped out of existence as an adventitious agent in rhesus monkey kidney cultures. There is no explanation for this, except as suggested below.

SV_4 and SV_{28} are very closely related agents with SV_{28} being the dominate member in respect to antigenicity. SV_{28} antiserum generally neutralizes SV_4 to about the same level as it does SV_{28}, but SV_4 antiserum, although neutralizing SV_{28}, generally does so, to 4-fold or lower titer than it does with the homologous virus. In effect, therefore, SV_4 could be considered a "prime strain" of SV_{28}, or a very slight serological variant of SV_{28}. Our more recent experience in virus recovery from rhesus monkey kidney cultures then would suggest that the dominant SV_{28} virus has replaced the "prime strain" (SV_4) which was so prevalent during the earlier period of observation.

SA_4 virus isolated from the African green monkey has received little attention in our laboratory. In some cross-neutralization tests it was found to cross one-way with SV_4, but not with SV_{28}. In tests with other antiserum preparations there has been no cross reactions. For further information concerning this virus the reader is referred to the papers by MALHERBE and HARWIN (83, 84).

Like the enteroviruses, these related types are quite stable at −70°C and at 4°C for long periods of time, and even survive 37°C for one month. They cannot be lyophilized.

IV. Simian Reoviruses

The simian reovirus group includes two viruses isolated from rhesus monkeys, SV_{12} and SV_{59} and one from the African green monkey, SA_3. Whether or not these viruses contain double stranded RNA is unknown, but since the human strains do, it is assumed that these simian varieties also do. They are fairly resistant to inactivation by formalin which would further suggest that their nucleic acid may be of the double stranded type. All three viruses produce the typical large globular type of acidophilic cytoplasmic inclusions as seen in cells infected with the human prototype, ECHO 10 virus (Fig. 4). In the original classification by CPE, SV_{12} and SV_{59} were placed in a common group with SV_4 and SV_{28} since the CPE produced by these agents was similar. A slight difference, however, was noted between the CPE of the SV_4 and SV_{28} pair, and that of SV_{12} and SV_{59}. The first time that ECHO 10 virus CPE was observed in our laboratory it was immediately recognized as being essentially identical to that which was produced by SV_{12} and SV_{59}. This lead to an extensive investigation into the possible serological relationship of SV_{12} and SV_{59} to ECHO 10 virus. This will be discussed subsequently.

SV_{12} has been isolated frequently from rhesus monkey kidney cultures and once from monkey stool (88). Of the 918 simian virus isolations made in our laboratory between April, 1955, and October, 1957, SV_{12} accounted for 173, or about 18 per cent. Its isolation rate has declined sharply in recent years. SV_{59} was originally isolated from lung tissue of a monkey with a respiratory illness by Miss *Nancy G. Rogers* (Department of Virus and Rickettsial Diseases, A.M.S.G.S., Walter Reed, Washington, D.C.). It was received in our laboratory labelled agent No. 59, and for this reason, was given the SV designation, SV_{59}, which resulted in the break in consistency of the numbering system. It has been recovered only one time from monkey kidney cultures in our laboratory.

The ultrafiltration studies reported by ATOYNATAN and HSIUNG (6) placed SV_{12} in the medium size range of 50 to 130 mµ which was in agreement with the size of other known reoviruses. I am unaware of any electron microscopic studies with this group of simian viruses. MINNER (94) in our laboratory found that SV_{12} would hemagglutinate chick, monkey, and human type 0 RBC's at 5°C and that SV_{59} agglutinated human RBC at 5°C. Both SV_{12} and SV_{59} were virulent for monkeys by intracerebral inoculation, but infection was limited to the epithelial cells of the chorioid plexus. This infection resulted in a flaccid type paralysis and death of the monkey. Identical findings were made in monkeys inoculated with ECHO 10 virus. Neither virus was pathogenic for small animals. Both SV_{12} and SV_{59} were found capable of growth in human cells.

The serological identification of these viruses and their differentiation from ECHO 10 virus by neutralization tests was difficult, especially when high titered antisera were employed. In an earlier report (56) we showed cross-neutralization between SV_{12} and SV_{59} in which the heterologous viruses were neutralized at a 4-fold lower titer than were the homologous viruses. Later when the similarity of these viruses to ECHO 10 was noted, and also that SA_3 isolated by MALHERBE and HARWIN (83, 85) was also found to belong in this group, more extensive tests were performed, not only in our laboratory, but in those of MALHERBE

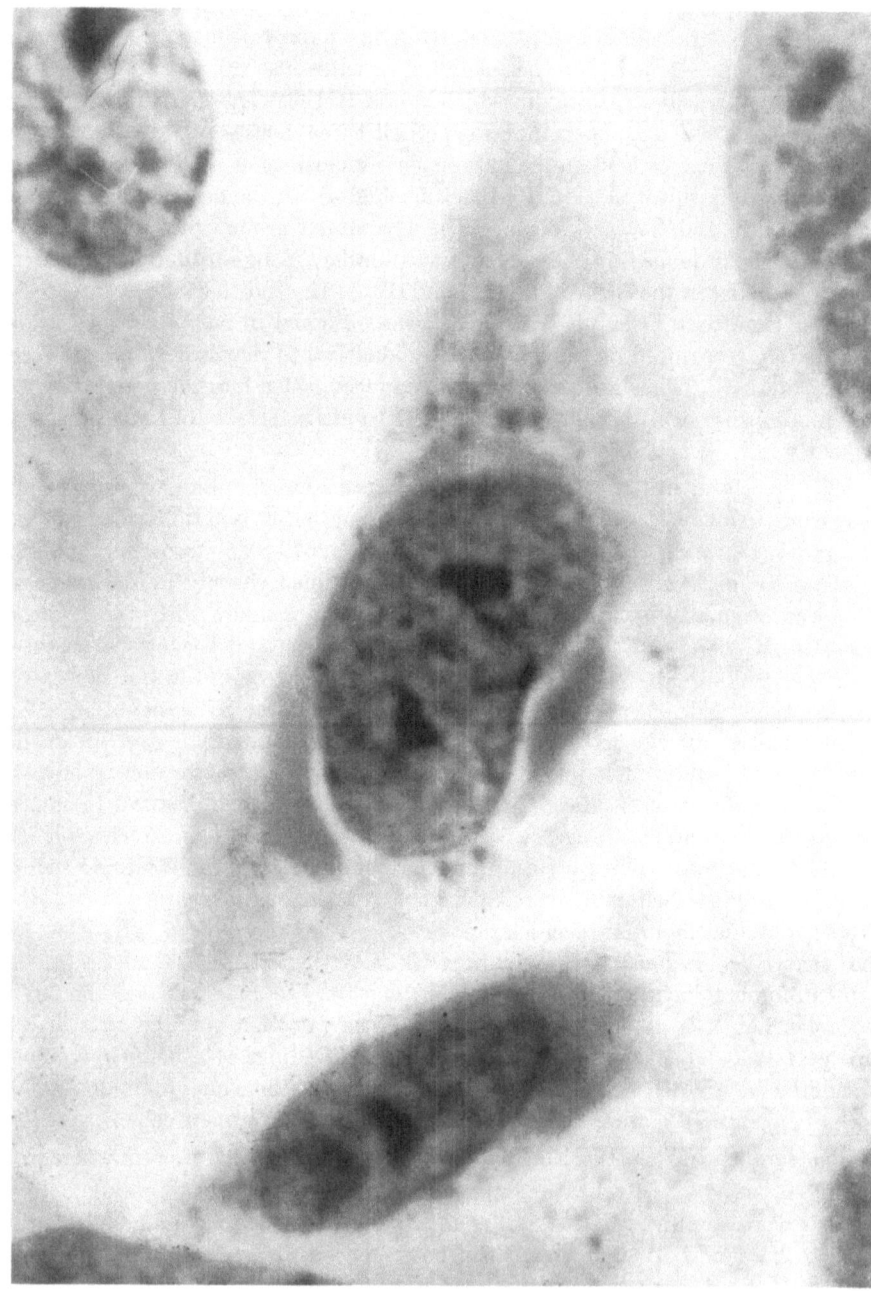

Fig. 4. Typical Reovirus cytoplasmic inclusion bodies produced by SA₃ in monkey kidney cells, hematoxylin and eosin stained. Approx. 1000 ×.

and WENNER. The general type of cross-reaction data obtained in our laboratory are seen in Table 10. We did not have antiserum to SA_3; thus, only one-way crosses with this virus are indicated. MALHERBE (84), however, did have all the antisera and the results obtained from his laboratory are seen in Table 11. WENNER (135) also performed numerous tests with this group of viruses, both by determination of serum endpoints, and by virus neutralization indices. His data from the latter type of assay are presented in Table 12. Considering the data in all three tables, the serological relationship between SV_{12} and SV_{59} is confirmed, and it appears also that SA_3 and ECHO 10 are more closely related to SV_{12} than to SV_{59}.

ROSEN (109) has studied SV_{12} and SV_{59} along with a group of other reoviruses by the hemagglutination-inhibition method. Based on his findings, he concludes that SV_{12} is a type 1 strain and that SV_{59} is a type 2. He also mentions in his text that SA_3 is a type 1 strain. Employing only antisera prepared against the human prototype strains of each of the 3 serological types of reoviruses, he has shown a greater distinction between SV_{12} and SV_{59} by HI tests than was possible by serum-neutralization tests. Certain of these data taken from Table 1 of his paper have been reassembled into Table 13. Since he did not include antisera prepared against SV_{12} and SV_{59} in his studies, these data do not show the complete cross-reaction possibilities by HI tests. His findings do, however, demonstrate the close relationship between ECHO 10 and SV_{12} and between SV_{59} and the D5 prototype strain of Group II reoviruses. With this human prototype virus antisera ROSEN showed no cross-reaction between SV_{12} and SV_{59}. ROSEN's data suggest that HI tests may be more specific for determination of serotypes amongst this group of viruses than were the neutralization tests employed in the other laboratories.

These simian reoviruses are quite stable, surviving $-70°C$ for years and $4°C$ for many months. They can also be preserved by lyophilization.

Table 10. *Cross-neutralization Data Obtained with Simian and Human Reoviruses*

Viruses	Antisera		
	ECHO 10	SV_{12}	SV_{59}
ECHO 10	*1024*	128	64
SV_{12}	256	*128*	32
SV_{59}	128	16	*128*
SA_3	64	16	16

Table 11. *Cross-neutralization Data Obtained with Simian and Human Reoviruses* (MALHERBE)

Viruses	Antisera			
	ECHO 10	SV_3	SV_{12}	SV_{59}
ECHO 10	*89*	20	25	10
SA_3	320	*562*	56	25
SV_{12}	47	117	*160*	11
SV_{59}	56	12	28	*80*

Table 12. *Virus Neutralization Indices for Simian and Human Reoviruses* (WENNER)

Sera	$TCID_{50}$ of viruses neutralized		
	ECHO 10	SV_{12}	SV_{95}
ECHO 10	100,000	20,000	320
SV_{12}	100	2,000	320
SV_{59}	10	320	1000

Table 13. *Typing of Human and Simian Reoviruses by Hemagglutination-inhibition* (ROSEN)

Viruses	Type	Sera		
		ECHO 10	D 5	Abney
ECHO 10	I	$≥ 1280$	<10	<10
D 5	II	<10	*40*	<10
Abney	III	<10	<10	*320*
SV_{12}	I	640	<10	<10
SV_{59}	II	<10	20	<10

V. Simian Herpesviruses

Four viruses have been isolated thus far from monkeys which can be classified as herpesviruses, based on their physical and chemical properties, their culture characteristics and the production of Type A intranuclear inclusion bodies in infected cells. B virus originally described by Sabin (117) is commonly isolated from rhesus monkeys and other Asiatic species. We have recovered the virus on several occasions from cynomolgus monkeys, as has been reported, also, by Wood and Shimada (136). There is a question, however, in these two laboratories as to the possible spread of the virus from rhesus monkeys to the cynomolgus, as both species were present. Hartley (39), on the other hand, has observed natural infection in cynomolgus monkeys which had not been exposed to rhesus monkeys. Endo et al. (29) have reported the presence of B virus antibodies in three Asiatic species, *Macaca fuscata, M. cyclopsis* and *M. iris*. Thus, it seems safe to conclude that B virus is indigenous in various species of Asiatic monkeys, but it has not been isolated, as yet, from African, or New World monkeys under natural conditions. That it can be artificially passed to African monkeys was demonstrated by Kirschstein (73). She successfully, although by accident, transmitted the virus from rhesus to African green monkeys, and patas monkeys by the oral route. In the same experiment she also infected cynomolgus monkeys.

Although the African green monkey does not appear to be a carrier of B virus, it does have a herpesvirus of its own, SA_8, isolated and described by Malherbe and Harwin (83, 85). In over six years' experience with African green monkeys we have never encountered, or isolated, SA_8 virus in our laboratory. It is likely, therefore, that SA_8 is not nearly so common a virus in the African green monkey as is B virus in the Asiatic species.

MHV, the marmoset herpesvirus was isolated from marmosets independently by Holmes et al. (45) and by Melnick et al. (91). Holmes et al. have suggested that this virus be called *Herpesvirus tamarinus* after the animal species from which it was first isolated. This virus is essentially 100 per cent fatal to the marmoset; thus, it seemed unlikely that this was the natural host of the virus. Further studies (20) have indicated that squirrel monkeys generally have high levels of antibody against this virus; thus, they may be the natural host. Antibody also was found in an occasional capuchin and spider monkey. The virus has been isolated also from owl monkeys by Hunt and Melendez (65).

SMV is a *temporary* designation for a new herpesvirus not yet described in the literature. It was isolated from a spider monkey by Lennette and has been further studied and characterized in our laboratory. Preliminary serological studies with different species of South American monkey sera suggest that the virus may be more or less restricted to spider monkeys in their native habitat, although an occasional squirrel or capuchin serum showed low levels of neutralizing antibody. This virus will be described in greater detail in a forthcoming publication from our laboratory.

Since these four simian herpesviruses are indigenous in monkeys from widely separated geographical areas it would not seem necessary to go into great detail as relates to the differentiation of the four types; however, as many laboratories now use, and frequently house together, monkeys of different species, it does become important to study isolates carefully in order to determine if cross-infection of species has occurred. The viruses can be distinguished by serological tests, but there are also some other differences within the group. All four simian strains, plus herpes simplex virus, grow readily in primary rabbit kidney cells, and in strain LLC-RK$_1$ (60). B virus, MHV and SMV all grow in human cells, but no information is available on the susceptibility of human cells to SA$_8$ virus. SMV, MHV, and herpes simplex virus will not grow in the rhesus monkey kidney strain, LLC-MK$_2$ (59), but B virus is readily cultivated in this strain. Again, no information is available concerning SA$_8$. All the viruses will infect African green monkey kidney cells. In rabbit kidney cells SA$_8$ and MHV are more destructive than are the other three viruses. The CPE produced by B virus in rabbit kidney is quite unlike that produced by the same virus in monkey kidney cells. In the latter, large multinucleated giant cells are seen with spike-like processes extending from the cytoplasm. It is a very striking CPE, which, in itself, is essentially adequate to identify the virus. The CPE in rabbit kidney, on the other hand, amounts essentially to rounding up of the cells without enlargement. The infected cells eventually undergo necrosis and slough off of the cell sheet. Herpes simplex virus tends to produce a CPE in rabbit kidney cells somewhat more like B virus does in monkey cells. The other simian herpesviruses produce CPE in rabbit kidney cells somewhat like that seen with B virus.

The virulence of these viruses for small animals also aids in their differentiation. B virus is highly virulent for the rabbit by all routes of inoculation. Herpes simplex virus, likewise, possesses considerable virulence for the rabbit, but requires larger doses by some routes of inoculation than does B virus. All five viruses will produce local skin lesions in rabbits, and in respect to SA$_8$, this is about the extent of its virulence. SMV produces local lesions which in some animals is followed by CNS involvement and death. MHV by the intracerebral route produces meningoencephalitis and death. MHV, SMV, and herpes simplex virus all are virulent for mice, while B virus infects only an occasional animal and the infection cannot be further passaged to other mice. All of these viruses will produce local lesions on some occasions following intradermal inoculation in guinea pigs. None produce generalized disease and death in this species.

What these viruses will do in respect to infecting other monkey species is little known, but such information is badly needed. As mentioned previously it has been shown that some African species can be infected experimentally with B virus but no generalized or paralytic type disease was evident. MHV, which probably is indigenous in squirrel monkeys without producing fatal disease is highly virulent for the marmoset. Beyond these observations our knowledge, relating to the susceptibility of various species of monkeys to these viruses, is a blank.

As already indicated, it is necessary to include herpes simplex virus in any comparative studies of this group of viruses since it is a common human virus closely related to this group of simian herpesviruses. Some cross-neutralization is seen amongst members of this group, but the degree of crossing will vary with different lots of typing sera. We also have noted some differences between rabbit and guinea pig antisera, but this may be due to the fact that guinea pigs generally produce higher titer sera than do rabbits immunized with these viruses.

Table 14 indicates the cross-reactions obtained in recent tests with guinea pig antisera. Some of these cross-reactions, especially those between SA$_8$ and B virus, are greater than those obtained in earlier studies, and are much more extensive than those obtained with lower titered rabbit produced antisera. SA$_8$ also shows cross-reactions with herpes simplex virus. The usual, one-way cross between B virus and herpes simplex virus is evident. MHV and SMV show only low level crossing with each other, and not with any of the other agents. NOWAKOWSKI and DEINHARDT (97) have shown an extensive one-way cross between these two viruses which we have yet to confirm in our laboratory.

The performance of neutralization tests with these viruses needs special comment. These viruses all require complement, or a complement-like accessory factor for neutralization. If fresh sera are employed satisfactory tests generally can be performed without "potentiating" the sera, but as a routine procedure, we add 5 per cent fresh normal guinea pig serum to the virus-serum mixture.

Table 14. *Cross-neutralization Data Obtained with Simian Herpesviruses*

Viruses	Sera				
	B-virus	H. simplex	MHV	SA$_8$	SMV
B Virus	64	<8	<8	32	<8
H Simplex	32	≥1024	<8	64	<8
MHV	<8	<8	≥2048	<8	8
SA$_8$	16	8	<8	≥256	<8
SMV	<8	<8	8	<8	128

This can best be accomplished by adding 10 per cent serum to the virus challenge dose, which, when mixed with equal parts of antiserum, reduces the guinea pig serum to 5 per cent. If sera are heat inactivated, then the addition of "potentiating" factor is a necessity to obtain the full potential titer of the antiserum. Low titered antisera (1:32 or less) when heat inactivated show no titer at all. If fresh normal guinea pig serum is added to such sera, the full titer is restored. The higher the titer of the serum, the lesser the effect of heat inactivation is noted.

The simian herpesviruses are quite labile viruses and will lose some titer on prolonged storage at −70° C. They survive for a few weeks at 4° C but with a considerable drop in titer. B virus is inactivated after 7 days' exposure at 37° C. We have not attempted lyophilization of these viruses; however, SABIN did lyophilize B virus successfully.

One agent, SA$_6$, is included in Group B listed under the herpesviruses (Table 2) as a representative of the simian cytomegaloviruses. This virus was isolated from African green monkey kidney cultures and described by MALHERBE and HARWIN (83, 85). We have been unable to cultivate the virus in our laboratories. The original investigators recovered a number of strains of this virus from kidney cultures, but found also that a similar agent (by histological studies) could be recovered from green monkey salivary glands. Focal lesions were produced in cultures, and by cytophagocytosis several cells were seen to combine. Eosinophilic inclusion bodies were seen in the nuclei of infected cells. Serological studies were limited due to the poor antigenicity of the virus. Thus, it is not known if all the strains isolated were identical. For very beautiful colored photographs of cells infected with this and other SA viruses, the reader is referred to reference (85).

VI. Poxviruses

Three viruses are included in this group but perhaps only one, monkey pox-virus, is a true poxvirus. Yaba virus has some poxvirus-like properties, while YLD is a newly isolated agent which produces a Yaba-like disease in monkeys. It has properties, however, which differ from those of Yaba virus.

Monkey poxvirus was first isolated and described by VON MAGNUS et al. (81). Two outbreaks of pox-like disease occurred in cynomolgus monkeys in their institute. The first outbreak lasted two weeks and involved 20 per cent of the monkeys in the group. In the second outbreak 30 per cent of the animals in the colony showed evidence of infection. The clinical disease in the monkey was characterized by a generalized petechial rash which progressed into a maculo-papular eruption. Lesions were seen over the entire body. The animals were not morbid, or otherwise affected by the disease except that scratching of the lesions suggested itching. The lesions gradually were covered with a crust which eventu-ally sloughed, leaving a distinct scar.

A virus was isolated from the dermal lesions which grew in monkey kidney, human amnion and HeLa cells. It was cultivated also in embryonated eggs and was virulent for rabbits, adult and suckling mice. The cultivated virus, on rein-oculation, produced dermal lesions in cynomolgus monkeys. Guinea pigs and chick-ens were resistant to infection. Electron micrographs of the virus obtained directly from the pustular lesions, and from the chorioallantoic membrane of egg passaged material, revealed the brick-shaped elementary bodies common to the pox-viruses with dimensions of approximately 200 by 150 mμ.

Serological studies revealed that the monkey poxvirus was closely related to vaccinia, but not identical to it. Hyperimmune rabbit and human vaccinia antisera extensively neutralized both viruses, although less monkey poxvirus was neutralized than vaccinia. In CF tests with this hyperimmune rabbit serum, complement fixation was about 2-fold lower with monkey poxvirus than it was with the homologous vaccinia antigen. Hyperimmune rabbit serum prepared against the monkey poxvirus fixed complement with vaccinia virus to consider-ably higher titer than it did with the homologous virus. Both vaccinia and mon-key poxvirus harvested from chick embryo chorio-allantoic membranes hem-agglutinated chicken RBC's. Tissue culture grown virus did not. In HI tests, 4HA units of vaccinia virus titered both vaccinia and monkey poxvirus antisera to the same level. An antigenic difference was demonstrated in agar-diffusion precipitation tests. Both viruses were tested against vaccinia antiserum. The precipitation patterns were divided into two major zones, each of which contained minor zones. The vaccinia virus produced three zones in one of the patterns while in the same area, monkey poxvirus produced only two zones of precipitation which were spaced further apart than the similar zones produced by vaccinia virus. These authors showed further that monkey poxvirus was serologically unrelated to herpes simplex or to B virus. In addition to the serological data, the similarity of monkey poxvirus to vaccinia was demonstrated by cross-protection tests in rabbits. Rabbits which recovered following scarification with either virus were immune to challenge with both viruses.

Monkey poxvirus produced lesions on the chorio-allantoic membranes of embryonated eggs and on the scarified rabbit eye very similar to those produced by variola virus. The fact, however, that it could be continuously passaged in adult mice by the intracerebral route, and in the rabbit by intradermal inoculation, served to distinguish the virus from variola. Further, no evidence was obtained that continued passage of the virus in mice or rabbits led to a transformation to vaccinia virus. The ability to produce variola-like pocks on the chorio-allantois was retained through three passages. The failure of the virus to infect mice by intradermal or intraperitoneal inoculation, along with its inability to agglutinate mouse red blood cells served to differentiate it from ectromelia virus. Attempts to identify the virus as cowpox by precipitation diffusion tests failed, and, also, the lesions produced on the chorio-allantoic membrane were unlike those produced by cowpox.

In embryonated eggs, the virus produced titers of 10^{-8} or greater, while in tissue culture the titers varied between 10^{-4} and 10^{-6}. Tissue culture fluids titered in eggs, however, were in the 10^{-6} to 10^{-7} range. The virus could be lyophilized and was stable for several months at $-60°$C. Considerable virus was lost during a week's storage at $-15°$C and a gradual decrease occurred over several weeks at $4°$C. The virus was quite resistant to inactivation by ether. Inactivation by a 1 in 4000 formaldehyde concentration at $37°$C was accomplished after 50 hours' exposure.

Various other reports of the isolation of monkey poxvirus have appeared in the literature, most of which confirmed or extended the studies of VON MAGNUS et al. SAUER et al. (119) and PRIER et al. (101) in 1960, described their experiences with outbreaks of monkey pox in laboratory colonies. They found the infection to be both more severe and more prevalent in *Macaca philippinesis* (cynomolgus) than in *M. mulatta* (rhesus). The more severe infections in *M. philippinesis* produced generalized edema, respiratory failure and some deaths. In one group of 18 *M. philippinesis*, all had HI antibody to monkey poxvirus, while only 6 in a group of 23 *M. mulatta* were positive. In a colony of 2000 monkeys composed of 56 per cent *M. mulatta*, 41 per cent *M. philippinesis*, and 3 per cent *Cercopithecus aethiops*, sporadic clinical disease occurred in about 10 per cent of the animals. Most of these cases were seen in *M. philippinesis* monkeys. These authors isolated and grew the virus in rabbit and monkey kidney cultures and in embryonated eggs. By cross-HI tests they were unable to distinguish the virus from vaccinia, but by cross-CF tests a slight serological difference was noted. They confirmed the virulence of the virus for the rabbit and for the mouse by intracerebral inoculation. They also produced lesions by inoculation of the guinea pig foot pad, but could not produce dermal lesions in this animal. Infected cells from monkey lesions or from tissue cultures, contained cytoplasmic inclusion bodies similar to those seen in vaccinia or variola infected cells. The lesions seen in the monkeys were described as similar to those seen in humans infected with smallpox.

Another interesting report by McCONNELL et al. (87) described an episode of monkey pox disease in cynomolgus monkeys following whole body X-irradiation.

These monkeys had been in the laboratory for some time, and were in apparent good health prior to X-irradiation. Forty-five days after a dose of 350 r one animal showed clinical evidence of pox disease, and a second animal did so four days later. Severe facial and cervical edema, hemorrhagic ulceration, dyspnea, and bloody diarrhea, depicted the severity of the disease. Both animals died 12 days after onset. The infection spread to one non X-ray treated monkey which survived. The virus was isolated and cultivated in rabbit and monkey kidney and in embryonated eggs. Like the other authors, they also found the virus to be identical with vaccinia by cross-HI tests, and to have virulence for adult mice by the intracerebral route. A survey of the colony of 28 cynomolgus monkeys revealed that 89 per cent possessed HI antibody to monkey poxvirus. One group of 45 cynomolgus monkeys in a separate area showed only 11 per cent positive for antibody. In contrast to some of the other reports, they found that 77.5 per cent of 67 rhesus monkey sera were antibody positive at levels of 1 : 80 or greater. They concluded that monkey poxvirus was a latent infection, activated by the X-ray treatment. VON MAGNUS et al. (81) reported also that they recovered the virus from kidney cultures prepared from monkeys in their colony at the time the outbreak of disease occurred. It may well be that monkey poxvirus is a latent virus in cynomolgus and rhesus monkeys, but, if so, the recovery of this virus from monkey kidney cultures is a rarity. It has never been isolated in our laboratory, nor identified in any of the many samples of simian viruses sent to us by other investigators. Based on our experience (unpublished) in the recovery of latent vaccinia virus from rabbits, the spleen is much more likely to contain virus than is the kidney. Perhaps if many monkey spleens were cultivated, more evidence of latent monkey poxvirus infection would be detected.

Yaba is a pox-like virus which produces histiocytomas in monkeys. The disease was first recognized by BEARCROFT and JAMIESON (9), in rhesus monkeys at Yaba, near Lagos in Nigeria. A tumor-like growth was first observed on the eyelid of a rhesus monkey. Within six months, 20 of 35 animals in the colony became infected. The infection spread naturally to one dog-faced baboon and was experimentally transmitted to cercopithecus monkeys. Patas and cercocebrus monkeys resisted the infection. The infected animals showed no evidence of morbidity or emaciation. Multiple tumors developed along the course of lymphatic drainage of the original tumor site. The growth eventually sloughed, and the area healed. No metastases to the viscera were noted. Histologically, the tumors were pleomorphic with a predominance of large cells with abundant cytoplasm. Acidophilic rounded, or irregular-shaped bodies, of 1 to 5 μ were seen in the cytoplasm.

The virus was studied in greater detail by NIVEN et al. (96). They found that cynomolgus and cercopithecus monkeys were susceptible, but that patas, cercocebrus and capucin monkeys were resistant. Mice, rabbits, and embryonated eggs also were resistant. The virus particles seen in the cytoplasm of infected cells possessed the morphological and cytochemical properties of vaccinia virus. Electron micrographic studies of tumor sections revealed changes similar to those seen in poxvirus infected cells, but they could show no immunological relationship to vaccinia or Orf viruses. Dense particles of 280 mμ size were seen in cytoplasm. Some evidence of virus growth with CPE was obtained in monkey kidney cultures, but not in HeLa or chick embryo tissue cultures. CPE in monkey kidney cultures consisted of cellular enlargement with intense granulation of the cytoplasm. In later passages multinucleated cells with highly vacuolated cytoplasm were seen. Following three passages in culture, the virus retained its ability to produce tumors in monkeys but lost this on further passage.

Gradocol membrane ultrafiltration studies revealed that the virus would pass 0.65 μ, but not 0.23 μ membranes. The virus was successfully lyophilized, and was stable for at least eight months. The elementary bodies were visible by Victoria blue stain, and shown to be DNA by acridine orange.

The growth and assay of the virus in tissue culture was further studied by LEVINTHAL and SHEIN (79) who employed immunofluorescent technics to follow the development of the virus. They obtained growth in both primary human and primary green monkey kidney cultures. The human cells produced less virus than did the monkey cells. Little virus was released into the culture fluids, and the authors described contiguous spread of the virus from cell to cell. CPE was seen only when massive doses of antigen were present in the cells as determined by immunofluorescent staining. The virus was carried through three passages in monkey kidney and six in the human cells. Growth was slow with long periods between subcultures.

YOHN et al. (137, 138, 139, 140) have undertaken extensive studies with Yaba virus, including investigation of its growth and assay properties, purification and chemical characterization. Information relative to the latter two subjects was very kindly supplied by Dr. YOHN (141). In their initial studies they obtained growth of the virus in LLC-MK$_2$, MA-10 (human embryonic kidney) and BS-C-1 cell strains. No growth was noted in bovine kidney or chick embryo cell cultures. BS-C-1 cells were the most sensitive for assay of the virus extracted from tumors and in this respect were equal to assay by monkey inoculation. BS-C-1 cells were used for further studies of the growth kinetics of the virus *in vitro*. Virus synthesis was followed sequentially with light and immunofluorescent microscopy, and by histochemical technics. The virus grew in foci of infections and spread contiguously from cell to cell. Cytoplasmic inclusions were apparent after 4 days of incubation. The complete synthetic cycle required 50—60 hours. CPE was observed in five days in heavily infected cells, but with 100 infectious units or less per culture, 10 to 14 days were required to detect CPE. The foci developed the characteristics of "microtumors", three to four cell layers thick. Mitotic cells were seen at the periphery of the microtumors, but not in the center of the lesion. In further studies the optimal method for Yaba virus assay in BS-C-1 cells was determined. Medium 199 containing 50 per cent bovine amniotic fluid with 2.0 mM calcium and 1.0 mM of magnesium at pH 7.0 was the most satisfactory diluting fluid. Adsorption at 25°C for 18 hours followed by incubation at 35°C gave the most reproducible assay results. Direct enumeration of foci of infection by microscopic examination or by immunofluorescence was employed to determine end-points. End-points were expressed as FFU (focus forming units).

The synthetic cycle in BS-C-1 cells of Yaba virus extracted from tumor tissue was unlike that of vaccinia, rabbit pox or Shope fibroma virus. The vaccinia infectious cycle was four to six times more rapid than was that of Yaba virus. A further characteristic which distinguished Yaba virus from vaccinia was in respect to its ready inactivation by ether. Vaccinia virus was relatively resistant to such inactivation. In this respect Yaba virus was said to be more like the myxoma poxvirus group.

The virus has been carried through eleven passages in BS-C-1 cells with titers in the range of 10^6 to 10^8 FFU/ml obtained on the fifth to eighth day after inoculation. Virus neutralization tests were accomplished by determining the per cent inhibition of FFU.

YOHN and co-workers (141) have gathered extensive data relating to the extraction, purification and chemical analysis of Yaba virus, as well as from studies of metabolic inhibitors, and surveys of simian and human populations for antibody to Yaba virus. Details of these findings will be published in the near future. Virus was purified by means of Genetron treatment, sedimentation in sucrose gradients, digestion with nucleases and by isopycnic banding in CsCl gradients. From these studies they found the buoyant density to be 1.291. DNA was extracted from purified virus and analyzed chemically for base ratios. The guanine plus cytosine ratio for the virus was in the order of 33.0 to 36.0 moles per cent. The tumor tissue DNA contained ratios of 44.2 to 45.3 which was essentially the same value obtained with monkey liver DNA. Determination of virus base ratios by melting point curves and isopycnic density gradient centrifugation methods yielded guanine plus cytosine values of 32.0 and 33.0 per cent.

GRACE et al. (36) have demonstrated the susceptibility of man to Yaba virus. Six volunteers were inoculated intradermally and all developed nodules at the site of injection within five to seven days. The nodules grew slowly to a size of 2 cm, then rapidly sloughed. Virus was readily recovered from these lesions. The histology of the nodules was similar to that described for the monkey tumors. Serial passage was accomplished in man, and CF antibodies developed in infected individuals. Human susceptibility had been first demonstrated through a laboratory accident. A worker was accidentally stuck in the web of the hand with a Yaba virus contaminated needle. A slight erythema appeared several days later, but quickly disappeared. Four months later a nodule appeared and grew to a size of 2 cm before it was excised. The histopathology was like that of a Yaba tumor and CF antibody was present in the serum.

Yaba-like-disease virus, YLD, was isolated by ESPANA (30) from monkeys during an outbreak of Yaba-like-disease at the National Center for Primate Biology, Davis, California. Ten different species of monkeys were in the compound but only the rhesus and cynomolgus monkeys became infected. About 40 per cent of the animals in gang cages became infected. The lesions were unlike those of vaccinia or smallpox. They were more tumor-like, but had soft crater-like centers. The lesions involved the epidermis like vaccinia or molluscum contagium, whereas, Yaba tumors do not. These tumor-like lesions did not exceed 25—40 mm in size, and, thus, were smaller than Yaba tumors. The lesions healed in about two weeks, and the monkeys showed no other clinical signs of illness. A virus was isolated from the monkeys' lesions in BS-C-1 cells. The virus also was cultivated in primary green monkey kidney cultures, human embryonic skin-muscle cultures, WI38, and human embryo kidney, but grew poorly in HeLa and rabbit embryo kidney cells. No growth was obtained in LLC-MK$_2$ or RK$_{13}$ cell strains. The virus grew in foci as does Yaba virus, but progressed more rapidly. CPE with cell lysis was observed in three to five days and titers were in the order of $10^{-4.0}/1.0$ ml. Inclusion bodies were prominent in the cytoplasm. Following 12 tissue culture passages, the virus was still infectious for monkeys. Mice, guinea pigs, rabbits, and embryonated eggs resisted infection with YLD virus. In cross-CF tests with specific anti-sera against Yaba and YLD, no cross-reactions were seen.

During the course of this outbreak, eleven cases of the disease occurred in animal handlers. Bites, scratches, or some trauma preceded the human infections. The skin lesions in man were similar to those seen in monkeys, but other clinical symptoms were also present. These were marked by severe lymphadenopathy and high fever. All patients survived and developed CF antibody to YLD virus. These sera, with one exception, however, did not fix complement with Yaba virus.

YOHN and his associates (141) also studied the YLD virus obtained from ESPANA. By electron microscopic studies they concluded that it was a poxvirus, and also found that cells infected with YLD showed immunofluorescence with Yaba virus antiserum. They agreed, however, that it grew more rapidly in tissue culture than did Yaba virus, and that it produced smaller cytoplasmic inclusion bodies. Yaba virus CF antibodies were detected in two of six monkey sera tested but in only one of forty-two sera from animal handlers. The one positive serum was from a patient with active lesions. Sera from six other individuals who had previously been infected with YLD virus failed to react at significant levels with Yaba CF antigen. Stained sections of YLD tumors failed to reveal typical cytoplasmic inclusions and the histology was unlike that seen in the Yaba virus-induced tumors. Grossly, however, the two tumors appeared similar. Thus, from what is now known, YLD virus appears to be related to Yaba, but has certain features which distinguishes it from the original Yaba strain.

VII. Myxoviruses

This group contains two true myxoviruses, SV_5 and SV_{41}, as well as the foamy viruses, which, like measles virus, resemble the myxoviruses. SV_5, a frequent contaminant was one of the earliest recognized latent viruses in simian kidney cultures. It, perhaps, has attracted more attention in the literature than any of the other simian viruses with the exception of SV_{40}. SV_5 is indigenous in Asiatic and African monkeys, but has been isolated from man (53, 121) and antibody to SV_5 has been demonstrated in human serum (7). Its similarity to mumps and to CA viruses (now called parainfluenza 2) was recognized early (55) and later it was clearly shown to be a member of the parainfluenza group by CHANOCK et al. (16). Its identity as a myxovirus is confirmed by the following properties:

1. Hemagglutination of erythrocytes.
2. Sensitivity of red cell receptor sites to RDE.
3. Ether sensitivity.
4. Growth in embryonated chicken eggs.
5. Size: 90—150 mμ.
6. Nucleic acid core contains RNA.
7. Virus matures at cell surface.

It was reported previously (55) that SV_5 showed a marked seasonal variation in its occurrence as a latent virus in rhesus monkey kidney cultures, with the peak in the fall months. CHANOCK et al. (16) reported similar observations. Our original findings, however, were based on the appearance of CPE in the infected cultures, but, later, when hemadsorption was employed, the virus was isolated at other times of the year when CPE was not evident. More recent experience gained through studies with the African green monkey revealed that in this species, SV_5 was most prevalent in the kidneys from early spring to late fall, but produced a detectable CPE much less often than it did in rhesus monkey kidney cultures. Cultures prepared from monkeys received from December to April were rarely found to contain SV_5.

The CPE produced by SV_5 was characterized by patches of fused or coalesced cells containing from several to many nuclei and resembled giant cells. These areas of coalesced, or fused cells, in monkey kidney cultures never were as large nor contained as many nuclei as did the syncytia produced by foamy viruses, respiratory syncytial or measles viruses. HOLMES and CHOPPIN (47) however,

Table 15. *Cross-neutralization Tests with* SV_5
Mumps and Parainfluenza Viruses

Viruses	Sera				
	Para 1	Para 2	Para 3	SV_5	Mumps
Parainfluenza type 1	*32*	<32	<32	<4	4
Parainfluenza type 2	<4	*512*	<32	4	<4
Parainfluenza type 3	<4	<32	*256*	<4	<4
SV_5	4	<32	<32	*1024*	32
Mumps	4	32	32	32	*512*

did observe large syncytia in BHK-21-F cultures infected with SV_5, but these cells produced little infectious virus. These authors also commented on the minimal cytopathic effects seen in monkey kidney cultures. This is true sometimes as the extent of CPE may range from none at all, to extensive degeneration of the cultures. As previously noted, the extent of the CPE varied at different times of the year when the virus occurred as a contaminant in rhesus monkey kidney cells. In the serial passage of the virus in primary rhesus monkey kidney cultures the extent of CPE varied from time to time. Whether these observations reflected differences in the cells, or in the virus seed, remains to be determined. When the virus is continuously propagated in strain $LLC\text{-}MK_2$, however, CPE is a constant property.

The serological relationship of SV_5 to the parainfluenza viruses and to mumps is seen in Table 15. Immune rabbit sera were employed in these studies, prepared from animals whose pre-immunization sera were negative for antibody to this group of agents. The parainfluenza 2 and 3 sera were toxic at levels below 1:32, thus, possible low levels of crossing with these two sera were undetected. It is seen from these data that SV_5 and mumps viruses crossed both ways at low levels. Mumps virus was also neutralized to some extent by the parainfluenza antisera. CHANOCK et al. (16), compared SV_5 to parainfluenza Type 2 virus by CF and HI technics. Their findings are presented in Table 16. As noted, they found a one way cross-reaction by CF tests, as the parainfluenza 2 antigen reacted with SV_5 antiserum, but SV_5 antigen did not fix complement with parainfluenza 2 antiserum. No cross-reactions were obtained in the HI tests.

Two viruses isolated from man and described in the literature under different names appear to be serologically identical with SV_5. The SA virus was isolated by SCHULTZ and HABEL (121) in 1951 from nasal washings of an individual with a respiratory infection. These washings were inoculated into embryonated eggs, and then sub-passaged by the intracerebral route into hamsters. Mild to severe illness occurred in hamsters, in some cases, progressing to spastic paralysis and death. The virus isolated possessed all the properties of a myxovirus and was described as a new member of the group. Later, CHANOCK et al. (16) stated, but without supportive data, that the SA virus was serologically identical to SV_5. If this is true, the virulence of the SA virus for hamsters is not common to the SV_5 prototype virus (strain 2105-2WR). DA virus reported by HSIUNG and ISACSON (53) was isolated from the post mortem blood of a patient with infectious hepatitis.

Table 16. *Serological Relationships between SV_5 and Parainfluenza 2 Virus by Complement Fixation and Hemagglutination-inhibition*

Sera	Viral antigens			
	CF		HI	
	SV_5	Parainfluenza 2	SV_5	Parainfluenza 2
SV_5 pre	<10	<10	<10	<10
SV_5 post	32	80	1280	<10
Para 2 pre	<10	<10	<10	<10
Para 2 post	<10	320	<10	640

The virus was isolated both in human and in monkey kidney cells, and was repeatedly isolated from the same specimen in human kidney cultures. No other organs or tissues yield virus in either culture system. It appeared unlikely, therefore, that the virus recovered was a latent agent in the monkey kidney cultures employed. These authors demonstrated, however, that the virus was serologically identical to SV_5.

Many isolations of SV_5 virus have been made in numerous laboratories, and from a variety of sources. There is but little evidence that antigenic difference exists amongst these isolates. One strain was isolated in our biological control laboratories which did show some slight serologic variation from the prototype, but this was only evident when cross-neutralization tests were performed. SV_5 antiserum neutralized this new strain to the same extent as the homologous virus, but antiserum prepared against the new isolate neutralized the prototype SV_5 strain to 4-fold, or less, titer than it did the homologous virus. Thus, if one had only SV_5 antiserum available for identification purposes, this variant would not be detected. The designation of SV_{5A} was given to the variant, and, although it has never been described in the literature, its existence has been known by people working in the field through communication with our laboratory, and as the result of discussions held during meetings of the N.I.H. Simian Virus Committee.

ESPMARK (31) detected a difference in strains of SV_5 isolated in his laboratory in respect to hemagglutination, egg inoculation, serology, and plaque morphology. His strains were isolated from uninoculated primary cultures of cynomolgus monkey kidney. One strain designated AV188 was more like the prototype strain of SV_5 than was the other, referred to as AV174. Strain AV188 hemagglutinated fowl cells at all temperatures, as does the prototype strain, but the AV174 strain agglutinated cells only at 4°C and eluted rapidly at 37°C. Strain AV188 also produced hemolysis at high concentration. Cells from which strain AV174 had been eluted could be agglutinated by strain AV188, which suggested that the two strains had different positions in the "receptor gradient". By HI tests anti-AV188 serum inhibited both strains to the same degree, while anti-AV174 inhibited AV188 at a 4-fold lower titer than it did its homologous virus. In this respect these two strains behaved very much like SV_5 and SV_{5A} did in cross-neutralization-tests. In plaque assays, in monkey kidney cultures, strain AV174 produced smaller plaques than did strain AV188. In attempts to cultivate the two viruses in embryonated eggs, strain AV174 grew readily by the allantoic route of inoculation, while strain AV188 required three amniotic passages before adaptation to the allantoic sac was attained. The latter finding again was in agreement with our experience with the prototype strain of SV_5. Although we have not compared SV_5 and SV_{5A} by all the methods employed by ESPMARK, the indications are, that SV_5 and strain AV188 are similar, as are also, strains AV174 and SV_{5A}.

CHOPPIN (18), CHOPPIN and STOEKENIUS (19), and HOLMES and CHOPPIN (47) have made a very thorough study of the structure and growth properties of SV_5. Their work was not done with the prototype strain, but with one referred to as W3 shown to be serologically identical to the prototype virus. By electron micrographic studies, CHOPPIN and STOEKENIUS (19) found the virus to be similar to the NDV, mumps and parainfluenza group. Most particles were spherical and approximately 120 mμ in diameter, but some pleomorphic particles measured up to 460 mμ in their largest dimension. The complete virions contained an internal component which was a flexible single stranded helix with an elliptical cross section. It appeared to be composed of ellipsoidal subunits with a long axis of 55 to 70 Å, and about 25 Å on the short axis. The diameter of the internal component was in the order of 150 to 180 Å. Projections, 100 Å in length were seen on the surface of the virus envelope. These findings were in keeping with similar observations made on other members of the NDV, mumps and parainfluenza group of viruses.

CHOPPIN (18) described a minimal CPE produced by his strain of SV_5 in monkey kidney cells, but which was similar to that previously described (55). Following inoculation of 40 to 70 PFU (plaque forming units) per cell there was a latent period of 6 to 7 hours. Exponential growth occurred between 12 and 30 hours with a doubling time of 50 minutes. Virus production continued for 31 days in such cultures. By 24 hours after inoculation, the virus concentration had increased to 500 to 1500 PFU/cell. When cultures were inoculated at high multiplicities of virus to cells, no autointerference, production of interferon, nor development of incomplete virus occurred. Cells heavily infected with SV_5 virus retained their sensitivity to Coxsackie, ECHO, poliovirus, vaccinia, VSV, and influenza viruses. Although the virus produced minimal levels of CPE in tube cultures it readily produced plaques under agar which were detected by staining with neutral red. The plaques were turbid and appeared to be lightly or irregularly stained as compared to the surrounding normal cell sheet. The cells were still intact, but appeared to have lost their ability to take up the vital dye. Plaques generally were 1.5 to 2.5 mm in diameter and appeared on the fourth day after inoculation. Peak titers were obtained at six or seven days. A linear relationship existed between the number of plaques counted and the virus concentration. In comparison to tube assays the ratio of PFU to $TCID_{50}$ was essentially unity. Adsorption of SV_5 to monkey kidney cells was 50 per cent complete after 30 minutes and 90 per cent after two hours. The latter period of adsorption was employed in the plaque assays.

Virus synthesis and its effect on host cells was studied further by HOLMES and CHOPPIN (47). BHK-21-F cell strain (80) and primary monkey kidney cultures were employed. As mentioned previously, SV_5 caused fusion of cells and formation of large syncytia in BHK-21-F cultures which later disintegrated, but only low levels of infectious viral progeny were produced. A 7-hour latent period elapsed following inoculation of 15 PFU/cell. Doubling time in the virus growth cycle was 60 minutes. Giant cell formation began at six hours and progressed to a single large syncyticum. With high levels of virus multiplicity, the syncytia appeared in one hour. Time-lapse photomicrography demonstrated that giant cells formed by fusion of infected cells and that some of these polykaryocytes divided. Synthesis of cellular RNA, DNA, and protein was not inhibited in monkey kidney cells and not in BHK-21-F cells until extensive fusion of cells had occurred. The complete virion in monkey kidney cells was assembled at the cell membrane and released by budding. By immunofluorescence, specific fluorescence was first observed in the perinuclear region at about 3 hours after infection. By 7 hours the antigens were seen in the cytoplasm of all cells in the monolayer. These fluorescent foci increased in size, and number, and were found throughout the cytoplasm by 24 hours. These findings were again in keeping with the placement of SV_5 in the NDV, mumps and parainfluenza virus group.

These authors concluded that the principal difference in the growth and the effect of SV_5 on these two cell systems was its effect on cell membranes. In support of this they summarized the following observations:

1. Inverse relationship between degrees of cell fusion and virus yield;

2. Kinetics of virus multiplication similar during early stages of growth in both cell systems;

3. In spite of the low yield of virus, cytoplasm of BHK-21-F cells appeared to contain much viral antigen;

4. BHK-21-F cells contained large aggregates of the helical internal component of SV_5 in their cytoplasm. The altered membranes seemed unable to incorporate the internal component into complete SV_5 virions.

The prototype strain of SV_5 has shown no virulence for small laboratory animals, or rhesus monkeys as determined by clinical illness. Antibody response, however, has suggested that inapparent infections may occur. CHANG and HSIUNG (17) reported that DA virus did produce inapparent infections in mice, hamsters, and monkeys. Hamsters and monkeys responded with higher antibody titers than did mice and were resistant to reinfection. The virus multiplied in the lungs and tracheas of mice and hamsters following intranasal inoculation and produced titers of 10^6 or greater/gm in hamster tissues and about 10^4/gm in mouse tissues. Mice of all ages were susceptible. A viremia occurred between the second and seventh days after injection, and virus was present in the lungs and trachea on days one through ten. After 11 days, the virus was recovered from kidney, liver, and brain tissue, and frequency of isolation increased between the sixteenth and twenty-first days. After 21 days no virus was recovered. Mice could also be infected by the intraperitoneal route, in which case, the urine was the most frequent site of virus recovery. No viremia occurred, but a few mice yielded virus from their lung and trachea tissues, the liver, and occasionally the kidneys. The virus was recovered also from brain tissues between 11 and 15 days after infection. Histopathologic changes were limited to lung tissues. Minimal perivascular and peribronchial edema was observed on the fifth and sixth days. Infiltration with large and small mononuclear and plasma cells occurred, but no intra-alveolar exudate was noted. Focal atelectases was seen on the seventh day. That mice could be infected by direct contact with the virus was demonstrated by housing normal young animals in cages previously occupied by infected mice. In such experiments approximately 10 per cent became infected.

Attempts to produce infection in rhesus monkeys with SV_5 were hindered by the high incidence of natural antibody in these animals. A group of 26 monkeys, some negative, and some with low antibody titers (1:10 to 1:20) were inoculated intranasally with DA virus. Three of 26 became infected as evidenced by virus recovery from the lungs and trachea. Virus was not isolated from blood, urine, liver, or kidney by direct tests but the kidneys from two monkeys did yield virus following trypsinization and growth in monolayers. There was no clinical symptoms in any of the three monkeys which became infected. Pre-antibody titers of 1:10 or greater protected the monkeys against infection. The natural transmission of the virus in monkeys was demonstrated by placing two infected animals in common housing with six antibody negative monkeys. Five of the six contact animals became infected as evidenced by virus recovery from their lungs and tracheas. Virus, again, could not be recovered from suspensions of kidney tissue, but the kidneys did yield virus when grown in tissue culture. They emphasize the late recovery of the virus from such cultures as the cultures required 35 to 65 days' incubation before the latent virus was detected. This was in accord with the usual experience in the isolation of SV_5 from kidney cultures as the virus is most frequently detected in cultures which have been held for long periods of time, such as in vaccine safety testing, or from second generation cultures.

Transmission of SV_5 and disease production in young baboons was reported by LARIN et al. (77). Fifty-one animals free of respiratory illness were selected, and housed two or three to a cage. All animals were bled to obtain serum for antibody determination and nasopharygeal swabs were taken for virus isolation studies. One animal had a titer of 1:7 against SV_5 virus, and no virus was isolated from any of the swabs. One baboon in each cage was inoculated intranasally with SV_5 virus. All inoculated animals yielded virus and 25 of 26 contact controls also showed evidence of infection. The one contact animal which remained free of infection was the one with the pre-inoculation antibody titer of 1:7. By 40 days, however, this titer increased to 1:128. The inoculated animals excreted virus 24 hours after inoculation and for long periods thereafter. The contact baboons first excreted virus between 5 and 21 days after the start of the experiment, but they excreted virus for a shorter period of time (8—14 days) than did the inoculated animals. There was no viremia in any of the animals and antibody response was weak, in the order of 1:8 to 1:64. Mild clinical evidence of respiratory disease was noted.

TRIBE (131) has described experience in the infection of patas and cynomolgus monkeys with SV_5 and has reported favorable results in suppressing the incidence of latent virus in the kidney through immunization with an inactivated SV_5 virus vaccine. Observation on the frequency of kidney contamination with SV_5 in his laboratory revealed that 6 per cent of 312 patas monkeys and 13 per cent of 147 cynomolgus monkeys were positive. Sera from about 85 per cent of the patas monkeys had HAI antibody in the range of 1:7 to 1:224. Most fell into the 1:7 to 1:14 range. All monkeys whose kidneys were infected at the time of sacrifice had titers greater than 1:56. Monkeys which had passed through the unit and were then placed in quarantine were found to have virus in their throats 28 days later. Thus, a 4 week quarantine period was inadequate to eliminate the virus from the colony. Antibody studies revealed that nearly all monkeys passing through the unit became infected before being put into quarantine. A small percentage of patas monkeys developed viremia and this persisted for long periods. The author suggested that possibly only these animals have virus in the kidneys. Cynomolgus monkeys were more susceptible than patas. He suggested further that the virus was transmitted by human contact during feeding and cage cleaning.

The vaccine was prepared from virus passed through patas monkey kidney and then inoculated into calf kidney which gave the highest virus yields (256 HAU/0.25 ml). The virus also could be cultivated in pig and dog kidney culture. The virus was inactivated with 10 per cent perchlorethylene and adsorbed to aluminium phosphate. One ml doses given intramuscularly to antibody free patas monkeys failed to stimulate detectable antibody levels, but a second dose given eight weeks later, produced titers to 1:900 eighteen days following the booster dose. Challenge experiments were done in which one-half of a group of patas monkeys were given a single dose of vaccine and 18 days later both groups were inoculated intramuscularly with live SV_5 virus. No detectable antibody was present at the time of challenge. Eighteen days following the challenge dose the vaccinees had HAI titers of 1:64 to 1:128, while the non-vaccinated group remained essentially negative. By 35 to 38 days the titers of the vaccinated group had increased by 4- to 6-fold, and at this point in time the control animals, now 18 days post challenge, had developed titers of 1:128. Viremia occurred in both groups three days after challenge, but terminated by the tenth day in the vaccinated group. The contact group still had virus in the bloodstream after 35 to 38 days. All animals were sacrificed, and their kidneys planted in tissue culture. None of the cultures prepared from the vaccinated monkeys were found to be infected with SV_5, but a 100 per cent incidence occurred in cultures prepared from the kidneys of the control group. Although a single dose of vaccine did not stimulate detectable antibody levels it did restrict the duration of the viremia and prevented the contamination of cultures prepared from the kidneys of such animals. In more extended experience, over a 5-month period, the kidneys from 143 vaccinated patas monkeys were found to be free of SV_5 contamination. During the same period the kidneys from 2 of 30 (6.6 per cent) non-vaccinated animals yielded SV_5. In cynomolgus monkeys none of 32 vaccinated animals whose kidneys were used for culture preparation were infected, while 6 of 47 (14.8 per cent) non-immunized monkeys did have virus in their kidney tissue. These results are most encouraging and suggest that other laboratories should investigate the possibility of vaccinating monkeys against SV_5. It would be even better if vaccine could be made available to the collecting companies, so that animals could be immunized immediately after trapping. One dose given at this time, followed by a second dose given upon arrival in the user's laboratory might eliminate the problem of SV_5 as a contaminant in monkey kidney cultures.

The laboratory infection of vervet monkeys with SV_5 was studied by HEATH et al. (41). The SV_5 prototype strain was employed. The animals were inoculated intranasally with 1.0 ml of virus. Antibody titers were followed by the HI test and virus isolation attempted from the nose, throat, and blood. Various levels of virus infection from 30 to 30,000 $TCID_{50}$ were studied. Infection resulted from all dosage levels, but was most consistent at the highest dose. Virus was excreted from the nose and throat for eight to nine days after inoculation, but no evidence of viremia was obtained. An antibody rise occurred in all inoculated animals, but, again, was greatest in those receiving the highest dose of virus. All animals had some HA inhibitor in their pre-inoculation blood, ranging from $1:10$ to $1:40$. Infected monkeys showed some clinical evidence of disease such as nasal discharge and inflammation of the pharyngeal mucous membranes. Some of the inoculated monkeys were rechallenged twice, at two and nine months after the first dose. The second dose produced a marked increase in antibody, but only two of seven monkeys shed virus from the nose and throat and for a period of only three days. The third infection produced a less dramatic response in the antibody titers and no virus was detected in the nose and throat following this dose.

SV_5 is a rather labile virus which can be preserved by freezing at $-70°C$ or greater, but some loss of infectivity may occur over prolonged storage. At $4°C$ the viability is lost in several weeks. Its half life at $37°C$ is one hour. The virus can also be preserved by lyophilization but some loss in titer is generally seen.

Before leaving the subject of SV_5 and contamination of primary monkey kidney culture with hemadsorbing agents one should note the paper by DOWDLE and ROBINSON (22) which describes non-viral induced hemadsorption in monkey kidney cultures. In their experience only 24 of 144 monkeys yielded SV_5 in cultures of their kidneys. On the other hand, 67 per cent were antibody positive at the time of sacrifice. Non-specific hemadsorption occurred, without accompanying virus isolation, which was related to the age, or condition, of the red cells employed in the test. Freshly prepared RBC's, regardless of the solution used for collecting, washing and suspension never produced non-specific hemadsorp-

tion. Cells stored for 24 hours at 37°C, 48 hours at room temperatures, or for 7
to 10 days at 4°C in any of several suspending solutions, did produce the non-
specific reaction. Those stored at 4°C first showed evidence of this after 72 hours
which increased on further storage. Fresh cells pre-treated with trypsin, potassium
periodate or RDE, also produced non-specific hemadsorption. This observation
was made in African green monkey kidney cultures as well as in those prepared
from rhesus cells, and also in cultures of rabbit, cat, and dog kidney. Several
continuous cell strains of monkey kidney origin also demonstrated non-specific
hemadsorption, but a number of human cell strains of non-kidney origin did not.
Thus, the phenomenon appeared to be a characteristic of kidney cells. Our own
experience concurs with their findings, as many times, hemadsorption has been
noted in primary monkey kidney cultures when a virus could not be isolated from
the supernatant fluids. This was a point of considerable concern until the report
by DOWDLE and ROBINSON explained this phenomenon.

SV_{41} is a myxovirus very similar to SV_5, but which differs primarily with respect to its animal virulence. The virus originally isolated from cynomolgus
monkey kidney cultures was described by MILLER et al. (93). Following their
initial isolation of the virus, 16 additional lots of cynomolgus kidney cultures
were found to be contaminated with this agent. We first encountered the virus
about a year later when it was recovered from the lumbar cord of a rhesus monkey employed in a measles virus neurotropic test. The monkey died 12 days after
inoculation, but without clinical evidence of disease. Histopathological examination revealed a diffuse, mild leukoencephalitis with very mild chronic, or sub-
acute meningitis in brain tissue, but both the cervical and lumbar cord sections
were negative except for traumatic changes at the site of inoculation in the
lumbar cord. During the succeeding year, SV_{41} was encountered in 12 different
plantings of monkey kidney cells, but then disappeared as it has not been isolated
at any time during the past four years.

MILLER et al. recognized the CPE produced by SV_{41} to be similar to that
seen in cultures infected with SV_5. The virus, however, could not be neutralized
with SV_5 antisera prepared in their laboratories nor with commercially available
SV_5 antiserum or with serum supplied from our laboratories. This was confirmed
by tests performed in our laboratory. The further study and characterization
of SV_{41} was done by MILLER and his associates, and the results are contained in
the paper referred to above. Its identity as a member of the myxovirus group was
confirmed by its sensitivity to ether, its growth in embryonated eggs, its ability
to hemagglutinate red blood cells, its size of 180—250 mμ, and its serological
relationship to SV_5 and parainfluenza type 2. It grew and produced a CPE in
primary cultures of cynomolgus and African green monkey kidney and chick
embryo cells. Growth was obtained also in primary bovine kidney, but without
a CPE. Titers in these systems were in the range of $10^{-6.5}$ to $10^{-7.5}/0.5$ ml. The
virus grew and produced CPE also in two continuous cell strains of human origin,
HeLa and HEp-2. A virus pool titered simultaneously in green monkey kidney
cells, embryonated eggs and weanling mice was found to have approximately
the same titer in each assay system. In our laboratory it was readily grown and
assayed in strain $LLC-MK_2$ cells.

SV_{41} possessed a surprising virulence for laboratory animals. By the intracerebral route it was infectious for mice, hamsters, guinea pigs, monkeys, and young chickens. Clinical signs of central nervous system disease was observed in all species. In the small animals this was marked by lack of co-ordination, involuntary movement, spastic paralysis, weight loss and ruffled hair. The monkeys showed muscular incoordination and generalized tremor. Histopathologic changes were evident only in the tissue of the central nervous system. Mice and hamsters had a mild lymphocytic meningoencephalitis with necrosis of the pyramidal neurons in the hippocampus. Sections from infected guinea pigs revealed a disseminated lymphocytic encephalitis and choriomeningitis, but with no neuronal involvement. Chickens developed a disseminated lymphocytic encephalitis with necrosis of the neurons. The monkeys which had been pre-treated with cortisone and inoculated with 0.5 ml of SV_{41} intraspinally, and 1.0 ml intramuscularly, in addition to the intracerebral inoculation, developed severe lymphocytic choriomeningitis, but without neuronal changes.

Serologically, the virus showed some cross-reaction with SV_5 and parainfluenza 2 viruses by CF tests, but not by hemagglutinin-inhibition, serum neutralization or hemadsorption-inhibition tests. Data obtained from the CF tests are seen in Table 17. It is noted that SV_{41} antiserum fixed complement to the same level with all three antigens and that some crossing occurred amongst all three viruses. Based on these findings, the authors suggested that SV_{41} might be the "master virus" in this group, due to its broader antigenicity. Tests with all other recognized myxoviruses failed to reveal any serological relation of any of these to SV_{41}. The virus was not neutralized by either human or monkey gamma globulin.

Table 17. *Complement-fixation Cross Reactions amongst SV_{41}, SV_5 and Parainfluenza Type 2 Virus*

Antigens	Sera		
	SV_{41}	SV_5	Para 2
SV_{41}	160	20	40
SV_5	160	320	10.
Para 2	160	40	320

SV_{41} produced hemagglutination of chicken RBC's only at 4°C, but agglutinated guinea pig cells equally as well at 4°C, 25°C, and 33°C. In respect to the chicken cell agglutination it behaved more like the AV174 variant of SV_5, as described by ESPMARK (31), than it did like the prototype SV_5.

As mentioned previously, SV_{41} was readily inactivated by ether. It was somewhat more resistant to inactivation with 1 to 4000 formaldehyde at 37°C, however, than was SV_5. SV_5 was inactivated under these conditions in less than 24 hours, but SV_{41} required 72 hours' exposure before titratable virus was lost. SV_{41} was found also to be more resistant to heat than was SV_5, as temperature of 56°C for one hour, while greatly reducing the titer, did not totally inactivate the virus. It was stable at 4°C for over one year. Preservation by freezing at —70°C resulted in loss of titer unless a stabilizer was used. Skim milk, lactose glutamate or glycerin provided stabilization of infectivity at low temperature.

In respect to its animal virulence, SV_{41} was more like the SA virus isolated by SCHULTZ and HABEL (121) than was SV_5. A comparative study of the SA virus and SV_{41} has not been reported, but perhaps should be investigated.

VIII. Myxo-like Viruses

This group of viruses is composed of those agents which have been referred to as "foamy viruses" due to the vacuolization which occurs in infected epithelial cultures. It probably would have been more appropriate to have called these agents "syncytial viruses" as the first, and most consistent cytopathic effect, is the formation of large giant cells, or syncytia, which sometimes develop vacuolization later in the infection. This gives the cell sheet a "foamy" or "lacy" appearance. The syncytia without vacuolization are a little difficult for the novice to see, or to recognize, in unstained preparations as they can be missed easily if cultures are rapidly scanned under the low power of the microscope. It is important that the light be not too bright, and that careful adjustment of the microscope focus be made so that different plains of focus in the cultures are viewed. Sometimes it is more or less what is not seen that suggest the presence of a foamy virus syncytium rather than what is seen. An otherwise healthy appearing cell sheet, which has areas in which individual cells cannot be seen, but which has some density, is likely to be infected with foamy virus. Careful focusing and adjustment of the light source will generally reveal that these areas are large giant cells, or syncytia, containing many nuclei. The nuclei in these syncytia are not necessarily arranged in rings as is common in measles virus infected cells, and there is no evidence of either cytoplasmic or intranuclear inclusion bodies. This latter feature, distinguishes foamy viruses from all other syncytial or giant cell producing viruses. They are very common contaminants in kidney cultures prepared from numerous species of monkeys. This will be considered in greater detail below.

What is now known as foamy virus was first recognized by ENDERS and PEEBLES in 1954 (28) when they encountered the agent in monkey kidney cultures used for the isolation and propagation of measles virus. Uninoculated control cultures were observed to have a CPE indistinguishable from measles virus in the unstained cultures, but virus recovered from such cultures did not fix complement with measles immune sera. In stained preparations it was observed that syncytia produced by foamy virus did not contain the typical cellular changes seen in measles virus infected cells. Measles virus produces an acidophilic inclusion-like body in the nucleus and margination of the chromatin. The new agent was further studied and proved to be a virus by RUSTIGIAN, JOHNSTON and REIHART (116). One of these authors, JOHNSTON, has continued to study this group of viruses over the intervening years and has become the authority on simian foamy viruses. These authors isolated a syncytial forming agent which they called MK-D. Subsequent isolates of similar agents were designated MK_1, MK_3, and MK_4. They demonstrated its filterability and that infectious material was bacteriologically sterile. In addition to monkey kidney cells they found that the virus could be propagated in HeLa cells, and human embryo kidney cells in which the typical CPE was produced. In fibroblastic cultures of monkeys testicular cells and human embryo skin-muscle cells degeneration occurred, but syncytia and vacuolization did not occur. They were unable to infect monkeys, rabbits, adult mice, one-day-old chickens and embryonated chicken eggs with the virus, but an occasional suckling mouse inoculated intracerebrally did become infected and died. Serial passage in suckling mouse brain was attempted, but could not be carried beyond

5 passages. Sera from normal rhesus monkeys, man, rabbits, mice, horses, chickens, cattle and pigs were screened for antibodies, but only the monkey sera were positive. Sixteen of 29 sera tested neutralized the virus. Immune sera to a number of recognized viruses were tested also, but none neutralized the foamy virus. The MK-D strain was sent to our laboratory where it was compared with some of the earlier isolated simian viruses (SV_1, SV_2, SV_4, and SV_5) and found not to be related to any of these agents.

Although we were able to cultivate the MK-D agent in primary rhesus monkey kidney cultures the "foamy viruses" were a source of frustration, and often aggravation, in our earlier studies in the isolation and characterization of latent viruses recovered from monkey kidney cultures. The foamy degeneration was frequently seen, but we were unable to obtain continued passage in order to characterize the agent. At one time, a few successful sub-passages were made in primary rhesus monkey kidney cultures and the temporary designation of SV_{13} (55) was given to the agent. It was lost, however, before the studies were completed, and as a result, the foamy viruses did not receive SV numbers. Possible reasons for these difficulties were defined later by PLUMMER (100). Foamy virus growth curve studies in monkey kidney cultures revealed that infectivity peaked sharply on day eleven, and then fell off, while marked CPE did not occur until the nineteenth day when the virus titer was quite low. Thus, by harvesting fluids at the peak of CPE, very low levels of infectious virus were obtained. The susceptibility of rabbit kidney cultures to foamy virus was reported by BROWN (13) and the use of this cell system has greatly facilitated the isolation and study of these agents.

During the mid and late 1950's numerous reports appeared in the literature describing the recovery of foamy viruses from various sources, their incidence in monkey colonies, etc. (15, 32, 43, 50, 51, 78, 114), but all of these probably involved the original serotype virus. JOHNSTON in 1961 (67) reported the isolation of a second serotype, as well as the recovery of virus from monkey throat swabs, and the incidence of the two types in different species. These studies were carried out in Taiwan, where 56 of 126 monkeys studied, yielded foamy viruses either from the throat, or kidney, or both. Throat isolates were made by inoculation of throat swabs into rabbit kidney cultures, while most of the kidney isolations were made by "unmasking" the virus through cultivation of the cells. In one experiment, kidneys from 13 of 22 monkeys were positive by the "unmasking" technic, whereas, ground tissue suspensions prepared from the same kidneys, and inoculated into previously grown cultures, yielded only two isolations. In comparing the rates of recovery from the two sites, 21 monkeys were examined. Seven had positive throat isolations, but were negative in respect to kidney tissue; three were kidney positive and throat negative; while eleven were positive in both. The virus was not only more prevalent in the throats of monkeys, but it was also present in higher titer. Virus was never isolated from urine, the urinary bladder, rectal swabs, or from other tissues. It was proposed, therefore, that the throat infection was the principle source of virus which spread through colonies of monkeys. Virus was recovered from the throats of infected animals continuously over a 10-week study period, but the monkeys never showed any signs of respiratory or other illnesses.

Of the 56 positive monkeys, 6 were cynomolgus, 50 were *M. cyclopsis*. Forty-nine of the strains were identified serologically, of which 35 were found to be identical to the original type 1 (FV_1) virus, while 14 were a new serotype, designated type 2 (FV_2). All 14 type 2 strains were isolated from Taiwan, *M. cyclopsis*. The cynomolgus monkeys yielded only type 1 virus. The viruses were widespread in different regions of Taiwan and were isolated from animals of all ages. The type 2 strain was identical to type 1 in respect to physical and biological characteristics.

Human sera were assayed for antibody to both types in order to determine if infection occurs in man. One-hundred sera from Taiwan natives living in areas with large monkey populations were tested. None were found to possess antibody to either virus. Five different pools of human gamma globulin, some obtained

Table 18. *Cross-neutralization Tests with Three Types of Foamy Viruses*

Virus		Rabbit antisera		
Strain	Type	Type 1	Type 2	Type 3
FV_{21}	1	1250	<4	4
FV_{50}	2	<4	360	<4
FV_{2014}	3	<4	<4	125

Table 19. *Cross-neutralization Tests with Four Strains of Foamy Virus*

Virus	Sera			
	FV_1	FV_2	FV 3	FV 4[1]
FV_1	512	<4	64	<4
FV_2	<4	256	<4	<4
FV_3	4	<4	1024	<4
FV_4	<4	<4	<4	64

[1] Monkey antiserum, all others are rabbit.

from Japan and some from the U.S.A. were also found to be negative for antibody to the two foamy viruses. There was no cross-reaction with measles virus immune sera. Antibody was frequently found, however, in monkey sera.

A third serotype of foamy virus was reported by STILES et al. (125). In a study of foamy virus isolates from rhesus and African green monkeys these authors recovered type 1 virus from kidney cultures of rhesus monkeys, type 2, plus a new serotype from the African green monkeys. The new isolate was serologically distinct from types 1 and 2 as seen in Table 18, but was similar in respect to other properties. In studying the distribution of these viruses in monkeys from different geographical areas, they found that 45 of 100 cultures prepared from North Indian rhesus monkeys contained type 1 virus. Thirty-six of 100 kidney cultures of East African cercopithecus monkeys were positive for virus. Twenty-one of these isolates proved to be type 3; 14 were type 2, and only one was a type 1 strain. This is in direct contrast to our own experience, as over a 7-year period, we have isolated only SA_1, which is identical to FV_1, from latent infections in green monkey kidney cultures. The frequency of isolation has been quite high in our laboratory, exceeding similar experiences with rhesus monkeys.

Type 4, FV_4, was isolated by JOHNSTON (68) from South American squirrel monkeys. Of 17 animals studied, three were positive for virus isolation from either the throat or kidney. The virus isolated could not be neutralized with antisera prepared against the known types. JOHNSTON has supplied cross-neutralization data as seen in Table 19. Although it was stated above that FV_3 was serologically distinct from FV_1 and FV_2, Table 18 does show a slight neutralization of FV_1 by anti-FV_3. JOHNSTON's data presented in Table 19 indicate a much stronger one-way cross, in that FV_3 serum neutralized FV_1 in dilutions of 1 to 64. The homologous titer of the FV_3 serum, however, was considerably higher than that of STILES et al. (125). JOHNSTON also found that FV_1 serum neutralized FV_3 at 1 to 4. The new type 4 isolate from squirrel monkeys showed no cross-reaction with any of the other serotypes. JOHNSTON has summarized the distribution of foamy virus serotypes in the various species as follows:

Serotype	Species from Which Isolated
FV_1	Rhesus, grivet, vervet, cynomolgus, *M. cyclopsis*
FV_2	Grivet, cynomolgus, *M. cyclopsis*
FV_3	Grivet
FV_4	Squirrel

Table 2, lists another foamy virus, FV_5, and there most likely is also an FV_6. JOHNSTON (68) had tentatively reserved FV_5 for a new serotype isolated from the throats of African bush babies, but it appears that an agent isolated from chimpanzees by ROGERS (108) may have been isolated first, in which case it would be given the FV_5 designation and that from the bush babies, FV_6. Neither of these two agents were neutralized by antisera prepared against the first four serotypes, but they have not, as yet, been compared with each other. For the time being, therefore, it will be necessary to indicate, only, that a fifth type of foamy virus has been encountered and that possibly a sixth type also exists. The isolation of these two agents from two additional species further points to the wide distribution of this type of virus in subhuman primates. The foamy viruses resemble both respiratory syncytial (RS) and measles viruses of man in respect to the CPE which they produce, but there is no evidence that these viruses produce any disease in monkeys resembling the illnesses produced by RS or measles virus in man.

The grouping of the foamy viruses in a myxo-like virus category is based on some properties which they share in common with the myxovirus group. They are ether and chloroform sensitive, and contain RNA in their nucleic acid core. Their size and structure has been described as myxovirus-like by JORDAN et al. (69). Four fresh isolates, one from rhesus monkey and three from green monkeys were studied by electron microscopic technics. The virions were found to be $100-300$ mμ in diameter, and to have protrusions on their surface membrane of about 10 mμ in length. The virus had an internal helical component,

$10-12$ mμ in diameter and more than 100 mμ in length. There was no "herring bone markings" and very little evidence of fine structure within the internal component. The lack of detail in the internal component was a specific characteristic of foamy virus and distinguished it from the myxoviruses. In respect to its structure, foamy virus was more like the NDV, mumps, parainfluenza group of myxoviruses than it was like influenza or fowl plague.

SA$_1$ virus originally isolated from vervet monkeys by MALHERBE and HARWIN (83) is also listed in the foamy virus group, but it is essentially identical to FV$_1$. It is included in the classification because it is frequently referred to in the literature.

Foamy viruses are rather heat labile as 60°C for 30 minutes completely inactivates the virus, and survival at 30°C is limited to two to three days. They can be stored at 4°C for several weeks without loss of titer, but will eventually be lost if held for prolonged periods at this temperature. Preservation at $-70°$C or lower is satisfactory and viability can be further stabilized by the addition of 10 per cent serum to the freezing medium. The viruses also can be stored in the lyophilized state.

IX. Papovaviruses

Although two viruses are listed under this category in Table 2, it is doubtful at this time that SA$_{12}$ belongs in such a group. This agent was isolated but one time by MALHERBE and HARWIN (85) from cultures of vervet monkey kidney. It produces a CPE in green monkey kidney cells similar to SV$_{40}$, but vacuolization is less prominent. Nuclear changes are seen in stained preparations which somewhat resemble those in SV$_{40}$ infected cells. Nuclear hypertrophy occurs, but the inclusions produced are more clearly defined than those produced by SV$_{40}$. Growth in green monkey kidney is slower than is that of SV$_{40}$, and there is no significant neutralization of the agent by SV$_{40}$ antiserum. We have not been able to definitely determine if the virus is of the DNA type, nor has it produced tumors in suckling hamsters during many months of observation. Thus, its inclusion in the group is based solely on some common cytopathogenic properties which it shares with SV$_{40}$.

The papovavirus group name was proposed by MELNICK (89) to include papilloma, polyoma and the vacuolating agent (SV$_{40}$), which, according to him, possess a number of common properties. The name was devised by using the first two letters of each virus in the group, i.e. pa-po-va. Papilloma virus includes both the rabbit virus and the human wart virus. Properties common to this group are:

Size: $40-50$ mμ

Capsomeres: 42

Buoyant density in CsCl: 1.30 g/ml

Double Stranded DNA

Absence of essential lipids

Thermal stability

Enhanced inactivation by divalent Cations

Slow growth, about 24 hours doubling time

Multiplication in nucleus with involvement of the nucleolus.

There are those who disagree with the similarity of some of these properties within the group, in particular, the size and number of capsomeres. GRANBOULAN et al. (37) have taken exception to this grouping, as they felt that more information was needed in respect to the structure of these viruses before they were placed in a common group.

SV_{40} is not only the most common contaminant of rhesus monkey kidney cultures, but is also the most talked about and most extensively investigated virus in the simian virus group. Literature reports dealing with investigations relating to SV_{40} virus exceed the combined reports on all of the other 56 classified agents. Many of these studies deal with work relating to the oncogenicity of the virus. It is neither the intent, nor within the scope of this review, to include an extensive coverage of such work. Remarks, therefore, will be more or less limited to various properties of the virus and to its significance to laboratory workers employing monkeys and monkey kidney cultures in their investigations.

The virus was first recognized by SWEET and HILLEMAN (127) when they attempted to pass rhesus or cynomolgus monkey kidney grown viruses into cultures of grivet monkey kidney. The viruses under study were neutralized with specific antiserum, but SV_{40}, present in the seeds as a contaminant broke through the serum and produced the characteristic CPE which led to its being dubbed "the vacuolating virus". They also recovered the virus from normal rhesus and cynomolgus monkey kidney culture. Eight strains were described in the original report, of which, strain 776 was designated the prototype. In a limited effort to determine the incidence of SV_{40} in rhesus and cynomolgus kidney cultures, ten different lots of cultures of each monkey species were examined. Each of these culture lots were prepared from the pooled kidneys of two to three monkeys. Seven of the 10 lots of rhesus monkey kidney cultures yielded SV_{40}, while only one of 10 lots of the cynomolgus cultures did so. Virus titers of $10^{-6.0}$ or higher were present in the rhesus kidney culture, but there was no evidence of CPE. No hemadsorption was seen in SV_{40} infected rhesus cells, and it was for these reasons, that this very common virus went undetected for so many years in spite of the extensive use of rhesus monkey kidney culture for virus research and vaccine production.

Doses of 1000 $TCID_{50}$ will usually elicit the first signs of CPE 3 to 4 days post inoculation. The outstanding cytopathic feature of SV_{40} in green monkey kidney cultures is the vacuolization of the cytoplasm. These vacuoles, however, are quite small as compared to those seen in foamy virus infected cells, and further SV_{40} does not produce giant cells or syncytia. Some cells in the cultures may show no vacuoles, others only a few, but as the infection advances, many cells are full of vacuoles and have a honeycomb-like appearance. The vacuoles seen in living cells are highly refractile and the smaller ones are difficult to distinguish from granules. In hematoxylin and eosin stained cultures the vacuoles appear as holes in the cytoplasm with intensely stained boundaries. The same appearance is seen in cultures stained with acridine orange and viewed by ultraviolet fluorescent microscopy. Thus, no stainable material has been detected in the vacuoles. As infection progresses toward the terminal stage, five to ten days, the cells show degenerative changes, reduction in size and aggregates form which detach from the glass. The CPE of SV_{40}, plus serological studies performed by SWEET and HILLEMAN (127) eliminated the identification of the vacuolating agent as one of the previously recognized latent viruses of rhesus or cynomolgus monkeys. This was confirmed in our laboratory, and by mutual agreement this new agent was given the designation of SV_{40}.

The high incidence of SV_{40} in cultures of rhesus monkey kidney has rendered them useless for virus vaccine production under existing regulations, as well as for any other purpose in which the presence of a second, or contaminating agent, would influence the results of the experiment. SWEET and HILLEMAN suggested about a 70 per cent incidence in rhesus monkeys when the kidneys of two or three animals were pooled. HAVENS (40) in our Biological Development Laboratories, found a 100 per cent incidence in 354 rhesus monkeys tested in groups of 10 to 26 for cell culture preparation. Five-hundred and sixty-nine cynomolgus monkeys from Vietnam, and 714 from Java, however, tested under similar conditions, were all negative for SV_{40}. These observations were made on the presence of virus in monolayer cultures as determined by sub-inoculation of fluids into green monkey kidney cultures. SV_{40} isolations from MAITLAND type cultures (82) of either species however, was much less frequent, and by experimentation it was shown that SV_{40} grew more slowly and to lower titer in these cultures. MAITLAND cultures of rhesus, cynomolgus, and green monkey kidney were seeded with SV_{40} virus. Fluids from inoculated cultures were assayed six and ten days later. Titer increase in green monkey kidney over 10 days ranged from 1 to 3 log units. Rhesus cultures either did not support growth, or in a few instances, increases as much as 4 log units were obtained. Cynomolgus cultures produced only one log of virus during ten days' growth. The inocula used in these cultures probably were much greater than would be present in latently infected kidneys (before placed in culture) and since, even under these conditions, poor growth and low yields of virus were obtained, it seemed likely that the infrequency of SV_{40} recovery from MAITLAND cultures of monkey kidney was the result of the low levels of virus present in the kidney tissue, plus the inadequacy of the MAITLAND type culture system to support growth of SV_{40} virus. Thus, the use of MAITLAND cultures prepared from cynomolgus monkey kidney virtually eliminated the SV_{40} problem in virus vaccines prepared in such cultures in our laboratories. Unfortunately, poliovirus is one of the few viruses which can be propagated in MAITLAND cultures for vaccine preparation.

Although the data presented above indicated that SV_{40} virus was not latent in cynomolgus monkeys, this was true only for those tested at that time. SWEET and HILLEMAN reported that 60 per cent of the cynomolgus monkeys which they tested were carrying SV_{40}, and the recovery from Indonesian cynomolgus monkeys was reported by ZEITLYONOK et al. (142). The recovery of SV_{40} from patas monkeys was reported by HSIUNG and GAYLORD (52). There are no reports, however, of its isolation from African grivet, or vervet monkeys which have not been in contact with cynomolgus or rhesus monkeys.

Further data relating to the incidence of SV_{40} in monkeys as they arrived in the United States, were reported by MEYER et al. (92). These determinations were made through antibody assays of sera from rhesus, cynomolgus and African green monkeys. Sixty-nine per cent of 58 rhesus sera were positive, 3 per cent of 39 cynomolgus sera, and none of 65 green monkey sera. They reported further, that arrivals with antibody were likely to yield virus positive kidney cultures, but that monkeys held in isolation for long periods of time, and which remained antibody free, did not have latent SV_{40} virus infection in kidney tissues. Although the green monkeys showed no evidence of natural infections, these authors demonstrated their susceptibility, both to experimental infection with SV_{40} virus, and through contact with infected animals. Monkeys inoculated by the intranasal route developed no signs or symptoms of illness, but did have a viremia, and shed virus in nasopharyngeal secretions and in the feces. Thus, African green monkeys should never be shipped, housed, or otherwise exposed to Asiatic monkeys.

SV_{40} can be cultivated in several cell systems, but generally produces the typical CPE only in green monkey, patas monkey, and baboon kidney cultures (52). The production of CPE and virus assay in established cell strains of green monkey kidney has been reported also. These included BS-C-1 (48), GMK, AH-1 (38), CV-1 (74), and $LLC-MK_3$ (64). Experience with both BS-C-1 and $LLC-MK_3$ strains has revealed a decreased sensitivity to SV_{40} in advanced culture passage levels. Thus, for best results it is desirable to maintain low passage frozen stocks which can be returned to as the passage level increases to the point where sensitivity to SV_{40} begins to decline. Growth does occur in rhesus and cynomolgus kidney cultures, but CPE is generally not seen. According to MEYER et al. (92) growth of SV_{40} occurred in the rhesus strain, $LLC-MK_2$, and some CPE without vacuolization was noted between 17 and 31 days after inoculation. We have been unable to obtain either growth or CPE in this strain in our laboratory. Growth without CPE has been obtained also in human kidney culture and some established human cell strains. Virus could not be cultivated in such non-primate cultures as rabbit, dog, swine, bovine, or duck kidney, nor in cultures of chick embryo cells (52).

Although most authors have reported that SV_{40} did not produce a cytopathic effect in rhesus kidney cultures, EASTON (24) has described such an observation in rhesus kidney cultures held for long periods of time. He reported that the virus grew slowly, but progressed to high titer, and after several weeks' incubation destroyed the cell sheet without producing vacuolization. Inocula of 5.5 log units added to cultures seven days after planting produced peak titers of $10^{9.0}$/ 6.0 ml of medium on the thirty-sixth day. Granulation of cytoplasm was seen and intranuclear changes occurred like those seen in green monkey kidney cells. EASTON (23) also studied the infection of rhesus cells by electron microscopy. In these cells the virus was seen only in the nuclei, whereas in green monkey kidney cells, virus was found in both the nuclei and cytoplasm. Naturally infected rhesus cells were observed to have $20-30$ mμ particles in the nuclei twelve days after planting when the titer of the supernatant fluid was $10^{7.3}$/ml. In experimentally infected rhesus cells as many as $10^{5.7}$ particles (by calculation) per nucleus were obtained 22 days after infection. No cytoplasmic particles were seen, and there was no evidence of CPE in the culture at this point in time.

The growth and changes produced in green monkey and baboon kidney cells by SV_{40} as observed by electron microscopy were reported by GRANBOULAN et al. (37). Thin sections of infected cells were prepared and examined periodically from three hours after infection to the eleventh day. At 3 to 6 hours slight nuclear hypertrophy was noted which became prominent at ten hours with involvement of the nucleoli and disturbance in the nucleolonema. Such changes increased in intensity until 48 hours when large numbers of virus particles were first seen in the nucleus and the homogenicity of the chromatin was lost. The nucleoli became smaller and contained dense areas of substance formed by the accumulation of finely granular material. The number of nuclei with virus particles gradually increased between days three and seven. Spherical bodies with clusters of virus particles were seen which appeared to be surrounded by a thin outer membrane. These structures were said to be similar to ones seen in polyoma infected cells. Crystalline aggregates of virus were also noted. The cytoplasmic vacuoles were evident at this point in time, and appeared as holes in the cytoplasm just as they did under the light microscope. Virus particles were seen in the cytoplasm by the sixth day. The vacuolization, however, was not regarded as the result of virus migrating to the cytoplasm as the authors felt that the cytoplasmic changes were a consequence of the nuclear pathology. By the tenth and eleventh days when cell lysis had started, the nuclei were densely packed with virus. The virus in these thin sections, embedded in Epon and double stained, were 33 mμ in size and showed a striking regularity in size. Particles seen in crystal-like arrangements were 40—44 mμ in diameter. The development of virus in baboon kidney cells paralleled these findings described for the green monkey kidney cell.

As previously mentioned, SV_{40} is an ether resistant double stranded DNA virus which is extremely stable at various temperatures, including 37°C, where it is stable for over two weeks. Although in their original report SWEET and HILLEMAN stated that SV_{40} was inactivated by 1 in 4000 formaldehyde at 37°C, and thus, was not a problem in inactivated virus vaccines, this has since been disproved. GERBER et al. (35) described a biphasic inactivation curve with 1 in 4000 formaldehyde at 37°C, in which a first order reaction was obtained during the first 50 hours of treatment, but the inactivation then leveled off, and one to two log units of virus were still detectable after fourteen days of exposure. Similar studies in our laboratories have confirmed the rapid inactivation of virus during the first three days of treatment, but in some instances viable virus was still detectable after 30 days of treatment. With concentrations of formaldehyde as great as 1 in 400, viable SV_{40} virus was still recovered after fourteen days, at which point the experiments were terminated. The virus was inactivated with a 1 in 500 concentration of BPL, but not with concentrations of 1 in 2000. Throughout these studies it was obvious that the initial titer of the virus was the important factor in respect to inactivation with formaldehyde. In low concentrations, of three log units or less, it was generally inactivated by 1 in 4000 formaldehyde within the thirteen-day period employed in the inactivation of poliovirus for vaccine preparation. Higher concentrations in the order of five logs, or more, could not be completely inactivated even under the extreme conditions mentioned. Thus, SV_{40} still presents a serious problem even to formaldehyde inactivated vaccines. Even if it could be reliably inactivated, regulations now do not permit the processing into vaccine of any virus harvests containing SV_{40} or other adventitious agents.

The high incidence of SV_{40} contamination in rhesus monkey kidney culture, plus its other peculiar biological and physical properties has created many problems for the vaccine manufacturer and virus researcher, but the most devastating finding was the demonstration of its oncogenicity for suckling hamsters. In 1961, EDDY et al. (25) reported that they had produced tumors in one to three-day old hamsters by the inoculation of extracts of rhesus monkey kidney cells. This report was disturbing to many people and it caused great concern in our laboratory in respect to whether or not our established monkey kidney cell strains possessed such properties. We, therefore, inoculated suckling hamsters with both primary rhesus monkey kidney cell extracts and similar ones prepared from strain LLC-MK_2 according to the technics described by EDDY et al., and since it was then known that SV_{40} was a common contaminate of primary rhesus cultures we also inoculated hamsters with SV_{40} virus. At the time of the second report by EDDY et al. (26) in which they defined the oncogenic property of the cell extract as being SV_{40} virus, we had tumors in the hamsters which were inoculated with SV_{40} virus, but not with extracts of either primary cells or of strain LLC-MK_2. EDDY et al. proved that SV_{40} was the causative agent in these experiments by recovering it from tumors produced by the cell extracts and by reproducing tumors upon its further passage into suckling hamsters. They also obtained tumors in 11 of 13 hamsters inoculated with the prototype SV_{40} virus. Many reports have since been published describing the oncogenicity of SV_{40} for suckling hamsters and many ramifications of this property.

EDDY (27) reported also that tumors were not produced in mice, guinea pigs, or rabbits following SV_{40} inoculation. KIRSCHSTEIN and GERBER (72) described ependymomas produced in newborn hamsters following intracerebral inoculation of SV_{40}. Similar lesions produced in an African rodent, *Rattus natalensis*, following subcutaneous inoculation were described by RABSON et al. (102).

The *in vitro* "transformation" of cells by SV_{40} also has been the subject of numerous reports. Such phenomena have been reported for cells of several species including man. SHEIN and ENDERS (123), for example, obtained "transformation" of human kidney cells and KOPROWSKI et al. (75) reported transformation of human tissue grown in organ culture. For additional information relating to the "transformation" of human, monkey, hamster, rabbit, mouse, porcine and bovine cells by SV_{40} *in vitro*, attention is directed to references (11, 12, 21, 66, 103, 104, 130). These represent but a few of such reports available in the literature.

The impact of the discovery of the latent SV_{40} virus on virus vaccine studies has been tremendous in spite of the early impressions of SWEET and HILLEMAN who stated: "In the practical sense, the vacuolating agent appears to be just 'one more' of the troublesome viruses to be conquered in the quest for vaccines which are safe and effective when used in man." This statement was made prior to the demonstration of the oncogenicity of SV_{40}, before its extreme resistance to formaldehyde was fully recognized, and before regulations were written which prohibited the processing, into vaccine, of any virus harvest found to contain adventitious agents. As SWEET and HILLEMAN pointed out, many vaccines, both live attenuated, and inactivated products produced in monolayer cultures of rhesus, or cynomolgus monkey kidney were found to be contaminated with SV_{40}. This was true also of most seed viruses. The continued supply of some vaccines was impaired and further clinical tests with new experimental immunizing agents were delayed due to the presence of this contaminant. In many instances new seeds had to be obtained and acceptable culture systems other than rhesus monkey kidney sought. Greatest difficulties were encountered in the case of the adenoviruses, which as demonstrated by BEARDMORE et al. (10) as well as others, would not grow in monkey kidney (rhesus or green monkey) unless SV_{40} was present. It was further shown, that on some occasions, adenovirus hybridized with SV_{40} (112) do yield a progeny with an adenovirus coat, containing an SV_{40} genome.

Many people were inoculated with vaccines containing live SV_{40} virus and many more ingested the virus along with experimental live, attenuated poliovirus vaccine. Although SV_{40} antibody developed following inoculation of the virus, no demonstrable acute illness was recognized nor was there any evidence for increased incidence of malignant disease in recipients within the limits of the observation made (33). Likewise, no ill effects were noted from ingestion of the virus, but MELNICK and STINEBAUGH (90) reported that some children became carriers of the virus and excreted it in the stool for periods up to five weeks. Experimental infections of man by the respiratory tract was demonstrated by MORRIS et al. (95). Intranasal installation of 10,000 $TCID_{50}$ produced a subclinical infection, but which was low grade based on the levels of antibody produced (1 to 5 to 1 to 80) and infrequency of recovery of virus from throat swab specimens. Twenty-two of 35 subjects given a mixture of respiratory syncytial (RS) virus and SV_{40} developed antibody to SV_{40}. Three of seven volunteers given the same material, but in which the RS virus had been neutralized by specific antiserum, yielded positive SV_{40} isolations from throat swabs taken 7 to 11 days after infection. That man probably becomes naturally infected by exposure to infected rhesus monkeys was demonstrated by SHAH (122). Sera collected from residents of Uttar Pradesh in Northern India, an area in which rhesus monkeys are prevalent, were assayed for SV_{40} antibody. Fourteen of 161 were positive at levels ranging from undiluted sera to 1 to 16 dilution. Ten of 37 workers engaged in the capture and care of monkeys also were found to possess SV_{40} antibody. None of these people had histories of ever receiving vaccines prepared in monkey kidney cultures. Thus, it is evident that man is susceptible to SV_{40}, but to date no ill effects have been encountered resulting from such infection. New regulations on vaccine production and the inclusion of additional testing procedures have eliminated the possible contamination of viral vaccines now in use with SV_{40}.

X. General Discussion and Considerations

Fifty-seven agents isolated from various species of sub-human primates were listed in the proposed classification presented in Table 2. This group represents agents which have been studied to sufficient degree to permit at least tentative classification into known, or recognized families, of viruses. Many additional agents have been recovered from monkeys, or from latent infections in tissue culture, which were not included in the table or in the text. A number of such agents were listed in Table 2 of reference 56. One virus, not considered in the discussion was MINIA, monkey-intra-nuclear-inclusion-agent, reported first by RUCKLE (113). It was later confirmed by RUCKLE (115) that this agent was immunologically and biologically identical to measles virus. The virus isolated from monkeys, undoubtedly was picked up through human contact, as it is well known that rhesus and cynomolgus monkeys become antibody positive for measles virus shortly after they are trapped and brought into close contact with man. The CPE produced by MINIA, like measles virus was similar to that produced by foamy virus. The intra-nuclear inclusion bodies, however, readily distinguished it from foamy virus in stained preparation. Unlike foamy virus, MINIA, or measles virus, does not persist as a latent infection in monkey kidneys as it disappears with the development of serum antibody titers. Thus, it would be encountered only during the early stages of infection.

The isolation of enteric viruses from baboons has been reported by FUENTES-MARINS et al. (34). KALTER (71) has isolated five adenoviruses and several enteroviruses from baboons which are not typeable with sera prepared against known human and simian strains. We have received from other investigators several strains of adenoviruses isolated from African green monkeys, only one of which was found to be identical with SA$_7$, the only previously described adenovirus isolated from this species. Likewise, we have on hand, a number of viruses sent to us by other laboratories which were isolated from chimpanzees. These have not yet been studied, or identification attempted with simian virus typing sera. None of these, however, could be identified with sera prepared against known human serotypes. PARKS (98) has mentioned the recovery of an adenovirus, SAB, from a baboon and another one, Agent No. 364, from the South American marmoset. Thus, many more viruses indigenous in lower primates exist, and a great deal more effort will be necessary before all of these can readily be identified and placed into a classification such as that proposed in Table 2. The increased interest in the use of sub-human primates, principally for cancer and chronic degenerative disease research has lead to the study and investigation of many types of primates in search for a suitable experimental animal for such studies. Virologists, some years ago, passed over the South American primates, primarily because they were resistant to infection with poliovirus, and other agents receiving intensive study at the time, and the use of African monkeys was minimal until after 1960. The rhesus and other Asiatic species not only were readily available in earlier years, but also met the needs of the virologists at the time.

Many laboratories now use and house monkeys of various species from wide geographical origins. Improper isolation of the species, either by the distributor, by the research laboratory, or by both, plus the close contact with man, has complicated the attempts to define the "natural viral flora" of a given species. The use of live virus vaccines in both the animals and the animal handler has added further to the confusion. KALTER (71) mentioned the spread of live, attenuated poliovirus vaccine from animal handlers to chimpanzees, and other larger primates. He stated further, from serological studies with gorillas, chimpanzees, orangutans, baboons, and monkeys, that these animals reflect in their antibody spectra their experiences in respect to association with man. Most viruses recovered from chimpanzees could be identified as recognized human viruses. A paper describing these studies is in press. Such experience with the larger species more closely related to man phylogenically, is in contrast to that gained through experience with rhesus and other species of monkeys. Throughout all of our studies of viruses isolated from monkeys, only once was a recognized human virus recovered and this was Asian influenza virus in the year 1957 when the virus was widespread throughout the world (see reference 56). Measles virus as mentioned previously, has been isolated also from monkeys. Just recently we identified an agent isolated by STANTON (Maj. Veterinary Corp., U.S. Army) from a gibbon with encephalitis. The virus was shown serologically to be Herpes simplex virus. In spite of these complications, it appears from data collected thus far, that for the most part, different species of primates, and those from different geographical areas, more or less, carry their own specific serological types of viruses. SV_5 and some types of foamy virus appear to be the exception to this. SV_5 is quite common in both Asiatic and African species, but thus far, has not been recovered, or reported, in South American species. FV_1 and FV_2 are also found in some species of both African and Asiatic monkeys, but FV_3 has been isolated only from African monkeys and FV_4 only from South American squirrel monkeys. SV_{40} is carried by several Asiatic monkeys, but thus far has been isolated from only one African monkey, *Erythrocebus patas*.

Although the classification of the simian viruses into recognized families of viruses has not proved difficult, the problem of nomenclature remains a serious one. For some purposes, a name which designates the type of virus, and distinguishes it as a simian, rather than a human virus, is adequate. Thus, the proposal by PEREIRA et al. (99) that the simian adenoviruses be placed in the adenovirus classification with M_1, M_2, M_3, etc., designations meet those demands. Many investigators, however, would like the name to indicate not only the type of virus, but further, which species of monkey it is native to. For this latter reason, we have declined to place new viruses isolated from African or South American monkeys into the SV series, since most interested investigators now know, or recognize, a virus with an SV designation, to be one of rhesus or cynomolgus origin. Likewise, the SA (simian agent) designation used by MALHERBE and HARWIN (85) denotes viruses isolated from African vervet monkeys. For the uninformed, however, neither SV nor SA carries any meaning other than virus of simian origin. If more appropriate names are selected for these viruses, and a new numbering system developed for serotypes such as in the proposal of PEREIRA et al. it will cause a great deal of confusion in the minds of investigators already familiar with these agents. It would be helpful, however, to the novice and to generations to come. This problem is under study by the N.I.H. Simian Virus Committee. (This committee, referred to frequently throughout these discussions, and in the reference list, was organized under the auspices of the National Cancer Institute, National Institutes of Health, Bethesda, Maryland. Dr. R. L. HEBERLING is the current chairman. The principal assignment of the committee is to collect information on simian viruses, establish prototype strains, and to arrange for the preparation and evaluation of standardized serological reagents to be made available for virus identifications.)

For the most part, the simian viruses are of little consequence to the general health of monkeys maintained in a colony. They constitute principally a nuisance factor to the investigators employing the animals in their research studies. A few exceptions to this have been noted. Several of the adenoviruses have been incremented in outbreaks of clinical disease, caused primarily by SV_{17} and SV_{32}. Some of the enteroviruses were isolated from monkeys with diarrhea, but it was never established that the virus was the etiological agent. B virus produces lip, buccal mucosa and eye lesions, but seldom causes death, or encephalitic disease in monkeys. The marmoset herpesvirus on the other hand, produces essentially a 100 per cent fatality rate when it invades a marmoset colony (46). As noted previously, however, this virus is probably indigenous in squirrel monkeys in which it produces little consequence. The new spider monkey herpesvirus (63) was isolated from an animal which developed crusty, brownish lesions on the lips and nose, and large, deep ulcerated areas on the tongue, palate and gums. Death occurred 24 hours later. The virus was recovered from brain tissue. There is no information, however, which indicates that the virus might occur in epidemic proportions. Yaba, Yaba-like disease, and monkey pox viruses produce disease in monkey colonies which spreads rapidly, but which is non-fatal, or otherwise of any great problem to the general health of the colony. The reoviruses, some of the adenoviruses and SV_{41} produce flaccid or spastic paralysis and death in some animals after intracerebral inoculation, but there is no evidence that such disease occurs under natural conditions.

The control, or eradication of latent viruses in monkeys, or in cultures prepared from their tissues, perhaps has not received the serious attention that it deserves. With either the Asiatic or African species commonly used, the control of SV_5, SV_{40}, and the foamy viruses most likely would eliminate in excess of 90 per cent of the infections. Our own experiences with African green monkeys over a seven-year period, revealed that SV_5 and SA_1 accounted for over 99 per cent of the latent virus isolations. SA_3 and SA_5 were isolated once or twice, but with one exception, no new, unidentifiable viruses were isolated. The exception was the appearance of a typical adenovirus type CPE in cultures held for long periods of time. This cytopathic effect, however, could not be sub-passaged to other primary cultures of green monkey kidney, established cell strain of rhesus or green monkey origin, or in a number of human cell strains. Similar findings have been made in other laboratories. Thus, in the African green monkey, only two latent viruses occur with any degree of frequency. SV_5 and foamy viruses are also quite prevalent in rhesus and cynomolgus monkeys, but in these species, they are second to SV_{40} as the principle problem. Although many other agents have been isolated from rhesus monkeys, the frequency of their occurrence was such that they would create but little problem, especially to laboratories employing relatively few monkeys.

Various approaches to the control of latent virus infections have been tried. Attempts to suppress SV_5 by the incorporation of high titered antiserum in the medium at the time of cell planting, and during virus assays, has been practiced by many laboratories and this method apparently does suppress the appearance of the virus for a while, but if cultures are held for long periods, three weeks or more, the virus eventually breaks through the antiserum. This method, however, cannot be used for vaccine manufacture where the virus must be grown in serum free medium. WALLIS and MELNICK (134) described the inhibition or suppression, of foamy virus, herpesviruses, and mycoplasma in cultures when 0.2 mM, aluminum chloride was included in the culture medium. It was shown further that no toxic or adverse effects occurred in respect to cell growth and maintenance, and that the sensitivity of cultures to a large number of other viruses was not impaired by the presence of aluminum chloride. Some laboratories hold frozen portions of the trypsinized cell preparation in liquid nitrogen, while the remainder of the preparation is planted in culture and observed for the appearance of latent agents. If none are noted, the frozen portion is then thawed, planted in culture and employed for whatever purpose was intended. Some loss in viability occurs during low temperature storage; thus, if this procedure is employed, the laboratory must be prepared to accept such loss. In our experience this amounts to about 20 per cent. If one-half of a cell suspension was frozen, while the other one-half was examined for latent virus, and found to be negative, this would provide fairly good evidence that SV_5, foamy virus and SV_{40} was not present in the preserved sample. These viruses, when they occur as latent infection, are generally present in sufficient titers to produce infection in all, or at least a large portion of the cultures, prepared from infected kidneys. Many of the other SV agents, however, including B virus, occur more sporadically, thus, testing of only one-half of the cell suspension would not necessarily provide assurance that the other half would be free of such agents.

Isolation, or quarantine of monkeys, is practiced in many laboratories and is a requirement for animals to be used in the production of live, attenuated virus vaccines. This is an effective measure if antibody negative (SV_5, SV_{40}, and foamy virus) animals are selected for the quarantine, and if they remain antibody negative during a period of approximately six weeks. Isolation, without serological tests, is of little value, since, as it has been pointed out in previous discussions, monkeys harbor some of these viruses for long periods of time, even though high titer, homotypic serum antibody is present. The isolation facility must be adequate to prevent the possibility of virus spread by direct contact, via air droplets, fomites, or animal caretakers. It is doubtful that many laboratories have isolation or quarantine quarters which provide such assurance.

The control of latent infections in monkeys by active immunization has been but slightly explored. As mentioned, and described earlier, Tribe (131) has attempted immunization with an inactivated SV_5 virus vaccine, which in his preliminary study, appeared to be highly effective in reducing the incidence of latent SV_5 infection in the kidneys of vaccinated animals. In view of our experience with the use of B virus vaccine, it was surprising that Tribe obtained apparent success in animals vaccinated so long after their initial capture, since by the time the animals reached the laboratory the infection should have been established in the group comprising the shipment. The B virus vaccine study referred to above (57), was initiated in India at the time the animals were first trapped. One-hundred monkeys were pre-bled and 50 were given the first dose of vaccine as they were taken out of the traps. The remainder were held as uninoculated controls. Ten per cent were antibody positive at the time of capture. A second dose was administered two weeks later when they were in our laboratories, and a third dose given 17 days after the second dose. The results can be very briefly summarized by pointing out that infection spread rapidly through the control group (and probably also the vaccinated group) during their first two weeks in captivity and did so probably before any effect of the vaccine was obtained. Vaccine response was difficult to asses due to the rapid development of natural antibody in the control group, but as best it could be determined, response was poor. This probably was due to the stress factors involved, since studies in other antibody negative monkeys which had been maintained in our laboratories for eight months before immunization, revealed good response to the vaccine. Animals with no detectable antibody developed titers in the order of 1 to 16 to 1 to 32 while others with low levels of pre-antibody obtained 4- to 32-fold increase in titer following immunization.

The susceptibility of man to infection with simian viruses is of considerable concern to all laboratories employing monkeys in their research programs. It also should be of concern to zoos, pet shops and to people who buy monkeys for pets. The greatest danger, of course, is readily recognized to be B virus, since this virus produces essentially a 100 per cent fatality rate in man. Various figures have been mentioned in the lay-press and in the scientific literature concerning the total number of human cases of B virus infection, but to the best of my knowledge, 18 is the number accounted for to date, and one of these is somewhat in question. Sixteen of these cases resulted in death. SOPER (124) prepared a review in August, 1959, which listed 15 cases reported to that date, and three additional ones have occurred in the meantime. Of particular interest is the fact that only four of the cases resulted directly from monkey bites, while in eight, the source of infection was undetermined. One case resulted from infection of a laceration received from a bottle containing monkey kidney cells, and one other was acquired while cleaning the skull of a monkey to be used as a curio. The fact that SABIN's original report (117) involved a case with monkey bite has lead people to believe that this is the chief source of infection, but as readily seen, neither a bite, nor any detectable wound may be necessary for infection to occur. Although the mortality rate is quite high in human infection, the morbidity rate appears to be low, and there is no evidence of sub-clinical infection. The low morbidity rate is evident from the fact that only 18 cases have been reported during the more than 30 years which have elapsed since the virus was first recognized. The extensive use of monkeys by industry during the past 12 years or more, and the relatively large numbers of people potentially exposed to infection through the use of monkeys, or monkey tissue cultures, plus the high incidence of B virus infection in rhesus monkeys, further attests to the low morbidity rate in man. Many bites, scratches, and other accidents have occurred without incidence. Several factors probably influence this low infection rate in man. First, the monkey involved presumably must have an active infection and be excreting virus. In well established colonies it is less likely that any given animal will be so infected than it is in a new group of animals just recently captured and shipped to the laboratory. Second, by antibody studies, we have found that about 50 per cent of the adult population has some B virus antibody, apparently as a result of Herpes simplex infection. Antibody to B virus produced in rabbits by hyperimmunization with Herpes simplex is fully as protective as is antibody developed against the homologous B virus. Negroes appear to have higher levels of Herpes simplex antibody than do Caucasians, and, thus, are more likely to have antibody to B virus. No case of B virus infection has been reported as yet in a negro. It should be stressed that Herpes simplex antibody, without the accompanying B virus antibodies, is ineffective. Third, in tissue culture, human cells are two to three log units less sensitive to B virus than are monkey or rabbit cells. The intact rabbit is only about 0.5 log units less sensitive than are tissue cultures prepared from rabbit kidney. Thus, one might project that man is relatively resistant to the virus, and that large doses might be necessary to elicit the infection even in an antibody negative individual.

In spite of the unlikelihood of any single accident resulting in a human infection, every precaution should be taken to avoid the possibility of infection. The first rule in any laboratory should be that rhesus, or other Asiatic species, will not be used unless they are paramount to the needs of the experiment. There may be equally as serious a problem with other species, but if so, these have not yet been defined. One possible B virus infection occurred in an individual bitten by an African green monkey (126). About one month prior to this accident, however, rhesus monkeys were introduced into the laboratory (8) and most likely were the source of the virus carried by the green monkey. If rhesus, or any monkeys for that matter, are used, every effort should be made to maintain a barrier between the monkey and the operator and no animal should be handled unless it is anaesthetized or heavily tranquillized. Visitors should be denied access to monkey storage areas, and personnel should refrain from playing with the animals or making pets of them. The strictest of personal hygiene and sanitation measures should be employed by all persons coming into contact with monkeys, monkey tissues, or items of equipment which are potentially contaminated. All primary monkey kidney tissue cultures, as well as all equipment used in the preparation and maintenance should be handled as though they were known to be contaminated with B virus. As a further precaution, but not as a substitute for the foregoing factors, active immunization of laboratory personnel against B virus infection would be a valuable adjunct. We have prepared an experimental B virus vaccine (57, 58, 62) which is highly effective in protecting rabbits against severe challenge with live virus, and which produces comparable antibody levels in man. This vaccine had been used for the immunization of approximately 300 persons in our laboratories during the past six years. Although no further cases of B virus infection have occurred in our laboratories since initiating the vaccine program, this does not constitute a measure of its effectiveness due first, to the low morbidity rate, and second, to the fact that far less rhesus monkeys have been used than in previous years. The efficacy of the vaccine in man probably can never be adequately demonstrated. A license application has been filed for this product, and hopefully, sometime in the future we will be able to distribute the vaccine to other laboratories who wish to use it.

Although several species of monkeys carry herpesviruses other than B virus, there is but one suggestion that any of these are virulent for man. This one case was referred to in an issue of the PRIMATE NEWSLETTER (120) and pertained to a report by N. W. KING, presented at the Annual Meeting of the Animal Care Panel, held in Chicago, Illinois; September, 1966. A research psychologist who had been in contact with squirrel monkeys, but who had received no known bites or scratches, developed an encephalitic disease. A rising titer to the marmoset herpesvirus *(herpesvirus tamarinus)* was noted in his convalescent serum aud provided circumstantial evidence that infection was due to this virus.

As recorded previously, Yaba virus will produce Yaba tumors in man follow-ing experimental, or accidental infection. No incidence of natural infection has been reported, however. The infection in man with Yaba virus appeared to be of little consequence as it was limited to local lesions which healed without sequellae. Transmission of the Yaba-like-disease virus to man does occur, but in the one instance reported, it was emphasized that bites, scratches, or some trauma pre-ceded the infections. Local nodules or tumors developed at the site of infection, and spread to other areas. A rather severe clinical syndrome followed, but all eleven patients survived the disease. Rhesus and cynomolgus monkeys were the source of the infection. Monkey pox virus has not been incremented in human infection, but it seems likely that man might be susceptible. The virus is so closely related to vaccinia, however, that man probably is immune as the result of smallpox immunization. Complete cross-protection in rabbits was demonstrated between monkey pox and vaccinia viruses.

The simian adenoviruses may be a potential source of infection and disease in man; however, only one frank infection has been observed. RAPOZA (106) reported an accidental eye infection produced by SV_{23}. In the course of animal inoculation, a plugged hypodermic needle resulted in a drop of virus being sprayed into the left eye of the investigator. Within 24 hours reddening of the eye appeared which increased in intensity with severe lacrimation. A full-blown conjunctivitis was present 48 hours after inoculation. Within a week, the right eye also became involved. The conjunctivitis persisted for five weeks. SV_{23} virus was recovered from the left eye four days after the infection. Serum obtained during, and after the episode, contained neither neutralizing nor HI antibody to SV_{23}. It appeared, therefore, that infection was limited to the eye. Recovery was complete without sequelae.

AULISIO, WONG, and MORRIS (7) reported an interesting situation in natives of New Guinea in respect to the incidence of antibody against SV_{20}. Sera from 267 infants, children, and adults living in two markedly different ecologic settings were assayed for antibody against this simian adenovirus. In the age group, 6 months to 2 years, 67 per cent were antibody positive, and the incidence in-creased with age to a peak of 82 per cent in the 15- to 18-year-old group. Antibody titers ranged from 1 to 10 to 1 to 160. In a similar survey of 42 sera collected in the United States, eight were found to contain antibody to SV_{20}, but these were in the age group from 10 to 18 years. Titers also were somewhat lower than in the New Guinea group. It appeared, therefore, that the New Guinea natives were experiencing an infection by a virus either identical to, or very closely related to SV_{20}, and that this experience increased with age. There are no naturally occur-ring primates in New Guinea other than man. We have been quite interested in these findings, since SV_{20} was one of the more highly oncogenic viruses in hamsters (61), and since a lymphoma-like disease occurs in the New Guinea po-pulation.

Antibody to SV_1 has been detected in human sera also. Of 80 sera collected from all ages between infancy and adulthood, we found ten to contain neutral-izing antibody against SV_1. Titers were generally low, in the order of 1 to 2 to 1 to 20. Only one of the positive sera was obtained from an infant. SV_1, and the other simian adenoviruses, react with many human sera in CF tests due to the common group antigen which they share with the human strains. A pool of human gamma globulin at a 1 to 10 dilution neutralized SV_1, SV_{20}, and SV_{25}, but none of the other viruses in the group through SV_{27}. The higher numbered simian adenoviruses have not been so tested. The significance of this antibody in human sera to simian adenoviruses is as yet unexplained. Since there was no cross-neutralization between the human and simian strains when specific antisera were tested, it would seem unlikely that the antibody found in human sera was the result of infection with human strains. AULISIO, WONG, and MORRIS' interest in testing human sera for SV_{20} antibody resulted from the apparent isolation of this virus from children with respiratory disease. It was never definitely determined, however, whether the virus came from the patients, or if it was present as a latent infection in the monkey kidney culture employed in the study.

That SV_{40} virus can infect man under experimental conditions by intranasal inoculation has been reported, and it also has been demonstrated that antibody is produced following parenteral inoculation. The natural infection of man by SV_{40} in India through close contact with rhesus monkeys has been determined also by antibody studies. No ill effects of such infections have been recorded to date.

SV_5 virus has been isolated from man on several occasions and antibody to this virus has been found in human sera. The significance of the latter observation, however, is questionable, due to cross-reactions of SV_5 with mumps and parainfluenza viruses. SV_5 is a very widespread virus and has been isolated from various sources where monkeys, or monkey tissue cultures were not present. We recovered it on one occasion from chick embryos and KRIM et al. (76) isolated it from a clone of human cells in a laboratory where there was no possible chance of contamination from a monkey source.

All of the viruses in the SV series with numbers below 30, plus SV_{59}, were studied for neutralization with human gamma globulin. Those which were neutralized included, SV_1, SV_2, SV_5, SV_6, SV_{12}, SV_{19}, SV_{20}, and SV_{25}. These findings with the three adenoviruses included in this group already have been discussed. The neutralization of SV_5 and SV_{12} is understandable due to their recognized relationship to the human parainfluenza and reoviruses. In this respect it was surprising that SV_{59}, a Type 2 reovirus, was not neutralized. The neutralization of the three enteroviruses, SV_2, SV_6, and SV_{19}, was unexpected, especially since relatively few monkey sera inhibit SV_2, although it is a common isolate from monkey stool specimens. The significance of these findings, again, is difficult to interpret in respect to the possible etiology of human infections.

The monkey has been a very valuable animal to medical scientists, as noted in the tribute by PHILLIP A. GOLAY, entitled, "Simian Sentiment". It is as follows:

> Where does the descent of man come in?
> For monkeys are monkeys and always have been.
> If they were our ancestors, as some folks say,
> How come the monkeys are with us today?

The whole tribe so intelligent and cute,

Sure did a smart thing to not evolute;

Give nothing but praise to the tail-swinging clan

In retaining their monkhood, regardless of man.

Thus, the monk, through the ages, by keeping his place,

Has proven a friend to the whole human race,

And really does something noble and big,

By giving his life as man's guinea pig.

We will praise Mr. Monk for giving us aid,

For through his blood our serums are made;

And our children are saved from dreaded disease,

By monks that live in their own family trees...

Mr. GOLAY has noted that the monkey has done "something noble and big, by giving his life as man's guinea pig". No one can dispute this contribution on behalf of "the tail-swinging clan", but we might enlarge upon Mr. GOLAY's sentiments with the following thought:

Monks of all species carry many viral agents

These frequently create problems as they remain latent;

But what would there be for scientists to do

If they didn't have such problems to pursue.

Acknowledgments

I would like to acknowledge the contributions of my previous co-authors in respect to the collection of data and observations pertaining to simian viruses presented in this report. This group includes: Mr. J. R. MINNER, Dr. J. W. SMITH, Dr. C. C. MASCOLI, and Mr. J. C. NASH. Further, I would like to thank the many people in the Biological Research, Development and Control Departments of Eli Lilly and Company, who have provided technical assistance, or who have shared their experiences with me, many of which have been incorporated into this review. Although this manuscript does not cover the entire subject of simian viruses, a much greater coverage was made possible through the very kind and helpful assistance of numerous investigators who supplied data, or information, through "personal communication", as recorded in the bibliography. Special acknowledgment and thanks are extended to these individuals. Last, but not least, I would like to express my gratitude to my secretary, Mrs. RUTH SUMMERS, who helped in the preparation of the manuscript, and who provided assistance in assembling and cross-checking the bibliography.

References

1. ANDREWS, C. H., F. M. BURNET, J. F. ENDERS, S. GARD, G. K. HIRST, M. M. KAPLAN, and V. M. ZHDANOV: Taxonomy of viruses infecting vertebrates: present knowledge and ignorance. Virology 15, 52—55 (1961).

2. ARCHETTI, I., D. STEVE-BOCCIARELLI, and G. TOSCHI: Structure of an enteric simian virus. Virology 13, 149—152 (1961).
3. ARCHETTI, I., and D. STEVE-BOCCIARELLI: Structure of some simian adenoviruses. Virology 20, 399—404 (1963).
4. ATCHINSON, R. W., B. C. CASTO, and W. McD. HAMMON: Adenovirus-associated defective virus particles. Science 149, 754—756 (1965).
5. ATCHINSON, R. W., B. C. CASTO, and W. McD. HAMMON: Electron microscopy of adenovirus-associated virus (AAV) in cell cultures. Virology 29, 353—357 (1966).
6. ATOYNATAN, T., and G. D. HSIUNG: Ultrafiltration of simian viruses. Proc. Soc. exp. Biol. (N.Y.) 116, 852—856 (1964).
7. AULISIO, C. G., D. C. WONG, and J. A. MORRIS: Neutralizing antibodies against simian viruses. SV_5 and SV_{20} in human sera. Proc. Soc. exp. Biol. (N.Y.) 117, 6—11 (1964).
8. BEALE, A. J.: Personal communication (1966).
9. BEARCROFT, W. G. C., and M. F. JAMIESON: An outbreak of subcutaneous tumors in rhesus monkeys. Nature (Lond.) 182, 195—196 (1958).
10. BEARDMORE, W. B., M. J. HAVLICK, A. SERAFINI, and I. W. McLEAN, Jr.: Interrelationship of adenovirus (Type 4) and papovavirus (SV_{40}) in monkey kidney cell cultures. J. Immunol. 95, 422—435 (1965).
11. BLACK, P. H., and W. P. ROWE: SV_{40} induced proliferation of tissue culture cells of rabbit, mouse, and porcine origin. Proc. Soc. exp. Biol. (N.Y.) 114, 721—727 (1963).
12. BLACK, P. H.: Transformation of mouse cell line 3T3 by SV_{40}: Dose response relationship and correlation with SV_{40} tumor antigen production. Virology 28, 760—763 (1966).
13. BROWN, L. V.: Pathogenicity for rabbit kidney cell cultures of certain agents derived from "normal" monkey kidney tissue. Amer. J. Hyg. 65, 189—209 (1957).
14. BULLOCK, G.: An association between adenoviruses isolated from simian tonsils and episodes of illness in captive monkeys. J. Hyg. (Lond.) 63, 383—387 (1965).
15. CARSKI, T. R.: A fluorescent antibody study of the simian foamy agent. J. Immunol. 84, 426—433 (1960).
16. CHANOCK, R. M., K. M. JOHNSON, M. K. COOK, D. C. WONG, and A. VARGOSKO: The hemadsorption technique, with special reference to the problem of naturally occurring simian parainfluenza virus. Amer. Rev. resp. Dis. 83, No. 2, pt. 2, 125—129 (1961).
17. CHANG, P. W., and G. D. HSIUNG: Experimental infection of parainfluenza virus Type 5 in mice, hamsters and monkeys. J. Immunol. 95, 591—601 (1965).
18. CHOPPIN, P. W.: Multiplication of a myxovirus (SV_5) with minimal cytopathic effects and without interference. Virology 23, 224—233 (1964).
19. CHOPPIN, P. W., and W. STOECKENIUS: The morphology of SV_5 virus. Virology 23, 195—202 (1964).
20. DEINHARDT, F.: Personal communication (1966).
21. DIDERHOLM, H., B. STENKVIST, J. PONTÉN, and T. WESSLÉN: Transformation of bovine cells in vitro after inoculation of simian virus 40 or its nucleic acid. Exp. Cell Res. 37, 452—459 (1965).
22. DOWDLE, W. R., and R. Q. ROBINSON: Non-specific hemadsorption by rhesus monkey kidney cells. Proc. Soc. exp. Biol. (N.Y.) 121, 193—198 (1966).
23. EASTON, J. M.: Electron microscopic observations of simian virus 40 in primary rhesus kidney tissue cultures. Proc. Soc. exp. Biol. (N.Y.) 114, 663—665 (1963).
24. EASTON, J. M.: Cytopathic effect of simian virus 40 on primary cell cultures of rhesus monkey kidney. J. Immunol. 93, 716—724 (1964).
25. EDDY, B. E., G. S. BORMAN, W. BERKELEY, and R. D. YOUNG: Tumors induced in hamsters by injection of rhesus monkey kidney cell extracts. Proc. Soc. exp. Biol. (N.Y.) 107, 191—197 (1961).

26. EDDY, B. E., G. S. BORMAN, G. GRUBBS, and R. D. YOUNG: Identification of the oncogenic substance in rhesus monkey kidney cell cultures as simian virus 40. Virology **17**, 65—75 (1962).

27. EDDY, B. E.: Tumors produced in hamsters by SV_{40}. Fed. Proc. **21**, 930—935 (1962).

28. ENDERS, J. F., and T. C. PEEBLES: Propagation in tissue cultures of cytopathogenic agent from patients with measles. Proc. Soc. exp. Biol. (N.Y.) **86**, 277—289 (1954).

29. ENDO, M., T. KAMIMURA, Y. AOYAMA, T. HAYASHIDA, T. KINJO, Y. ONO, S. KOTERA, K. SUZUKI, Y. TAJIMA et K. ANDO: Etude du virus B au Japon. Jap. J. exp. Med. **30**, 227—233, 385—392 (1960).

30. ESPANA, C.: Reported in N.I.H. Simian Virus Committee Meeting, March, 1967.

31. ESPMARK, J. A.: Some biological differences between two strains of simian parainfluenza virus — SV_5. Personal communication (1965).

32. FALKE, D.: Beobachtungen über das Auftreten von Riesen- und Schaumzellen in normalen Affennierenzellkulturen (Observation on the incidence of giant and foam cells in normal monkey kidney cell cultures). Zbl. Bakt. I. Abt. Orig. **170**, 377—387 (1958).

33. FRAUMENI, J. F., Jr., F. EDERER, and R. W. MILLER: An evaluation of the carcinogenicity of simian virus 40 in man. J. Amer. med. Ass. **185**, 713—718 (1963).

34. FUENTES-MARINS, R., A. R. RODRIGUEZ, S. S. KALTER, A. HELLMAN, and R. A. CRANDALL: Isolation of enteroviruses from the "normal" baboon (papio doguera). J. Bact. **85**, 1045—1050 (1963).

35. GERBER, P., G. A. HOTTLE, and R. E. GRUBBS: Inactivation of vacuolating virus (SV_{40}) by formaldehyde. Proc. Soc. exp. Biol. (N.Y.) **108**, 205—209 (1961).

36. GRACE, J. T., Jr., E. A. MIRAUD, S. J. MILLIAN, and R. S. METZGAR: Experimental studies of human tumors. Fed. Proc. **21**, suppl. 32—36 (1962).

37. GRANBOULAN, N., P. TOURNIER, R. WICKER, and W. BERNHARD: An electron microscope study of the development of SV_{40} virus. J. Cell Biol. **17**, 423—441 (1963).

38. GÜNALP, A.: Growth and cytopathic effect of rubella virus in a line of green monkey kidney cells. Proc. Soc. exp. Biol. (N.Y.) **118**, 85—90 (1965).

39. HARTLEY, E. G.: Naturally-occurring B virus infection in cynomolgus monkeys. Vet. Rec. **76**, 555 (1964).

40. HAVENS, M. L.: Personal communication (1961).

41. HEATH, R. B., I. E. FALAKY, J. E. STARK, R. H. HERBST-LAIER, and N. M. LARIN: Induced upper respiratory infection in the vervet monkey: a model of acute viral respiratory disease in man. Brit. J. exp. Path. **47**, 93—98 (1966).

42. HEBERLING, R. L., and F. S. CHEEVER: Some characteristics of the simian enteroviruses. Amer. J. Epidem. **81**, 106—123 (1965).

43. HENLE, G., and F. DEINHARDT: Propagation and primary isolation of mumps virus in tissue culture. Proc. Soc. exp. Biol. (N.Y.) **89**, 556—560 (1955).

44. HOFFERT, W. R., M. E. BATES, and F. S. CHEEVER: Studies of enteric viruses of simian origin. Amer. J. Hyg. **68**, 15—30 (1958).

45. HOLMES, A. W., R. G. CALDWELL, R. E. DEDMON, and F. DEINHARDT: Isolation and characterization of a new herpesvirus. J. Immunol. **92**, 602—610 (1964).

46. HOLMES, A. W., J. A. DEVINE, E. NOWAKOWSKI, and F. DEINHARDT: The epidemiology of a herpesvirus infection of New World monkeys. J. Immunol. **96**, 668—671 (1966).

47. HOLMES, K. V., and P. W. CHOPPIN: On the role of the response of the cell membrane in determining virus virulence. Contrasting effects of the parainfluenza virus SV_5 in two cell types. Plates 33 to 38. J. exp. Med. **124**, 501—520 (1966).

48. HOPPS, H. E., B. C. BERNHEIM, A. NISALAK, J. HIN TJIO, and J. E. SMADEL: Biologic characteristics of a continuous kidney cell line derived from the African green monkey. J. Immunol. **91**, 416—424 (1963).

49. HORNE, R. W., S. BRENNER, A. P. WATERSON, and P. WILDY: The icosahedral form of an adenovirus. J. molec. Biol. **1**, 84—85 (1959).

50. HOTTA, S., and C. A. EVANS: Cultivation of mouse-adapted dengue virus (Type 1) in rhesus monkey kidney culture. J. infect. Dis. **98**, 88—97 (1956).

51. HSIUNG, G. D., A. MANNINI, and J. L. MELNICK: Plaque assay of measles virus on *Erythrocebus patas* monkey kidney monolayers. Proc. Sox. exp. Biol. (N.Y.) **98**, 68—70 (1958).

52. HSIUNG, G. D., and W. H. GAYLORD: The vacuolating virus of monkeys. I. Isolation, growth characteristics, and inclusion body formation. J. exp. Med. **114**, 975—986 (1961).

53. HSIUNG, G. D., P. ISACSON, and R. W. McCOLLUM: Studies of a Myxovirus isolated from human blood. I. Isolation and properties. J. Immunol. **88**, 284—290 (1962).

54. HULL, R. N., J. R. MINNER, and J. W. SMITH: New viral agents recovered from tissue cultures of monkey kidney cells. I. Origin and properties of cytopathogenic agents SV_1, SV_2, SV_4, SV_5, SV_6, SV_{11}, SV_{12}, and SV_{15}. Amer. J. Hyg. **63**, 204—215 (1956).

55. HULL, R. N., and J. R. MINNER: New viral agents recovered from tissue cultures of monkey kidney cells. II. Problems of isolation and identification. Ann. N.Y. Acad. Sci. **67**, 413—423 (1957).

56. HULL, R. N., J. R. MINNER, and C. C. MASCOLI: New viral agents recovered from tissue cultures of monkey kidney cells. III. Recovery of additional agents both in cultures of monkey tissues and directly from tissues and excreta. Amer. J. Hyg. **68**, 31—44 (1958).

57. HULL, R. N., and J. C. NASH: Immunization against B virus infection. I. Preparation of an experimental vaccine. Amer. J. Hyg. **71**, 15—18 (1960).

58. HULL, R. N., F. B. PECK, Jr., T. G. WARD, and J. C. NASH: Immunization against B virus infection. II. Further laboratory and clinical studies with an experimental vaccine. Amer. J. Hyg. **76**, 239—251 (1962).

59. HULL, R. N., W. R. CHERRY, and O. J. TRITCH: Growth characteristics of monkey kidney cell strains LLC-MK₁, LLC-MK₂, and LLC-MK₂ (NCTC-3196) and their utility in virus research. J. exp. Med. **115**, 903—918 (1962).

60. HULL, R. N., A. C. DWYER, W. R. CHERRY, and O. J. TRITCH: Development and characteristics of the rabbit kidney cell strain, LLC-RK₁ (30044). Proc. Soc. exp. Biol. (N.Y.) **118**, 1054—1059 (1965).

61. HULL, R. N., I. S. JOHNSON, C. G. CULBERTSON, C. B. REIMER, and H. F. WRIGHT: Oncogenicity of the simian adenoviruses. Science **150**, 1044—1046 (1965).

62. HULL, R. N., and F. B. PECK, Jr.: Vaccination against herpesvirus infections. First PAHO-WHO international conference on vaccines against viral and rickettsial diseases of man. Washington, D.C. PAHO Scientific publication **147**, 266—275 (1967).

63. HULL, R. N., A. C. DWYER, and E. H. LENNETTE: To be published.

64. HULL, R. N.: Unpublished data (1961).

65. HUNT, R. D., and L. V. MELENDEZ: Spontaneous Herpes-T infection in the owl monkey *(Aotus trivirgatus)*. Pathologia Veterinaria **3**, No. 1 (1966).

66. JENSEN, F., H. KOPROWSKI, and J. A. PONTÉN: Rapid transformation of human fibroblast cultures by simian virus 40. Proc. nat. Acad. Sci. (Wash.) **50**, 343—348 (1963).

67. JOHNSTON, P. B.: A second immunologic type of simian foamy virus: Monkey throat infections and unmasking by both types. J. infect. Dis. **109**, 1—9 (1961).

68. JOHNSTON, P. B.: Simian Virus Committee Meeting. Personal communication (1967).

69. JORDAN, L. E., G. PLUMMER, and H. D. MAYOR: The fine structure of foamy virus. Virology **25**, 156—159 (1965).

70. KALTER, S. S.: Animal "orphan" enteroviruses. Bull. Wld Hlth Org. **22**, 319—337 (1960).

71. KALTER, S. S.: Reported at N.I.H. Simian Virus Committee Meeting, March, 1967.

72. KIRSCHSTEIN, R. L., and P. GERBER: Ependymomas produced after intracerebral inoculation of SV_{40} into newborn hamsters. Nature (Lond.) **195**, 299—300 (1962).

73. KIRSCHSTEIN, R. L.: Personal communication (1967).

74. KIT, S., D. R. DUBBS, P. M. FREARSON, and J. L. MELNICK: Enzyme induction in SV_{40}-infected green monkey kidney cultures. Virology **29**, 69—83 (1966).

75. KOPROWSKI, H., J. A. PONTÉN, F. JENSEN, R. G. RAVDIN, P. MOORHEAD, and E. SAKSELA: Transformation of cultures of human tissue infected with simian virus, SV_{40}. J. cell comp. Physiol. **59**, 281—292 (1962).

76. KRIM, M., S. C. WONG, and E. D. KILBOURNE: Isolation of a virus from a clone of a human cell line in continuous cultivation. Fed. Proc. **20**, 442 (1961).

77. LARIN, N. M., R. H. HERBST-LAIER, M. P. COPPING, and R. B. M. WENHAM: Controlled investigations in baboons *(Papio cynocephalus)* on transmission of SV_5 virus by contact. Nature (Lond.) **213**, 827—828 (1967).

78. LÉPINE, P., et M. PACCAUD: Contribution a l'étude du virus spumeux (foamy virus). 1. Etude des souches FV I, FV II, FV III isolées de cultures de cellules rénales de cynocéphales. Ann. Inst. Pasteur **92**, 289—300 (1957).

79. LEVINTHAL, J. M., and H. M. SHEIN: Propagation of a simian tumor agent (Yaba virus) in cultures of human and simian renal cells as detected by immuno-fluorescence. Virology **23**, 268—270 (1964).

80. MACPHERSON, I. A., and M. G. P. STOKES: Polyoma transformation of hamster cell clones — an investigation of genetic factors affecting cell competence. Virology **16**, 147—161 (1962).

81. MAGNUS, P. VON, E. K. ANDERSON, K. B. PETERSEN, and A. BIRCH-ANDERSON: A pox-like disease in cynomolgus monkeys. Acta path. microbiol. scand. **46**, 156—176 (1959).

82. MAITLAND, H. B., and M. C. MAITLAND: Cultivation of Vaccinia virus without tissue culture. Lancet **2**, 596—597 (1928).

83. MALHERBE, H., and R. HARWIN: Seven viruses isolated from vervet monkey. Brit. J. exp. Path. **38**, 539—541 (1957).

84. MALHERBE, H.: Personal communication (1958).

85. MALHERBE, H., and R. HARWIN: The cytopathic effects of vervet monkey viruses. S. Afr. med. J. **37**, 407—411 (1963).

86. MAYOR, H. D., R. M. JAMISON, L. E. JORDAN, and J. L. MELNICK: Structure and composition of a small particle prepared from a simian adenovirus. J. Bact. **90**, 235—242 (1965).

87. MCCONNELL, S. J., Y. F. HERMAN, D. E. MATTSON, and L. ERICKSON: Monkey pox disease in irradiated cynomolgus monkeys. Nature (Lond.) **195**, 1128—1129 (1962).

88. MELNICK, J. L.: Quoted at the DiCennial Conference on tissue culture. J. nat. Cancer Inst. **19**, 746 (1957).

89. MELNICK, J. L.: Papovavirus group. Science **138**, 1128—1130 (1962).

90. MELNICK, J. L., and S. STINEBAUGH: Excretion of vacuolating SV_{40} virus (Papovavirus group) after ingestion as contaminant of oral poliovaccine. Proc. Soc. exp. Biol. (N.Y.) **109**, 965—968 (1962).

91. MELNICK, J. L., M. MIDULLA, I. WIMBERLY, J. G. BARRERA-ORO, and B. M. LEVY: A new member of the herpesvirus group isolated from South American marmosets. J. Immunol. **92**, 596—601 (1964).

92. MEYER, H. M., Jr., H. E. HOPPS, N. G. ROGERS, B. E. BROOK, B. C. BERNHEIM, W. P. JONES, A. NISALAK, and R. G. DOUGLAS: Studies on simian virus 40. J. Immunol. **88**, 796—806 (1966).

93. MILLER, R. H., A. R. PURSELL, F. E. MITCHELL, and K. M. JOHNSON: A newly discovered myxovirus (SV_{41}) isolated from cell cultures of cynomolgus monkey kidney. Amer. J. Hyg. **80**, 365—376 (1964).

94. MINNER, J. R.: Lilly Research Laboratories unpublished data (1955).

95. MORRIS, J. A., K. M. JOHNSON, C. G. AULISIO, R. M. CHANOCK, and V. KNIGHT: Clinical and serologic responses in volunteers given vacuolating virus (SV_{40}) by respiratory route. Proc. Soc. exp. Biol. (N.Y.) **108**, 56—59 (1961).

96. NIVEN, J. S., J. A. ARMSTRONG, C. H. ANDREWES, H. G. PEREIRA, and R. C. VALENTINE: Subcutaneous "growths" in monkeys produced by a poxvirus (Plates I—IX). J. Path. Bact. **81**, 1—14 (1961).

97. NOWAKOWSKI, E., and F. DEINHARDT: Personal communication (1966).

98. PARKS, W.: N.I.H. Simian Virus Committee Meeting. Quoted studies performed by BENYESH-MELNICK, M. R. MIRKOVIC, C. F. PHILLIPS, and M. PARKS at Baylor University College of Medicine (1966).

99. PEREIRA, H. G., R. J. HUEBNER, H. S. GINSBERG, and J. VAN DER VEEN: A short description of the adenovirus group. Virology **20**, 613—620 (1963).

100. PLUMMER, G.: Foamy virus of monkeys. J. gen. Microbiol. **29**, 703—709 (1962).

101. PRIER, J. E., R. M. SAUER, R. G. MALSBERGER, and J. M. SILLAMAN: Studies on a pox disease of monkeys. II. Isolation of the etiologic agent. Amer. J. vet. Res. **21**, 381—384 (1960).

102. RABSON, A. S.: Ependymomas produced in *Rattus* (mastomys) *natalensis* inoculated with vacuolating virus (SV_{40}). J. nat. Cancer Inst. **29**, 765—787 (1962).

103. RABSON, A. S., and R. L. KIRSCHSTEIN: Induction of malignancy *in vitro* in newborn hamster kidney tissue infected with simian vacuolating virus (SV_{40}). Proc. Soc. exp. Biol. (N.Y.) **111**, 323—328 (1962).

104. RABSON, A. S., R. L. KIRSCHSTEIN, and F. Y. LEGALLAIS: Autologous implantation of rhesus monkey cells "transformed" *in vitro* by simian virus 40. J. nat. Cancer Inst. **35**, 981—991 (1965).

105. RAPOZA, N. P., and F. S. CHEEVER: Classification and identification of simian adenoviruses by the hemagglutination, hemagglutination-inhibition techniques. Fed. Proc. **25**, Abstract 340 (1966).

106. RAPOZA, N. P.: Personal communication (1967).

107. REIMER, C. B.: Lilly Research Laboratories. Personal communication (1964).

108. ROGER, N. M.: Reported at N.I.H. Simian Virus Committee Meeting (1967).

109. ROSEN, L.: Serologic grouping of reovirus by hemagglutination-inhibition. Amer. J. Hyg. **71**, 242—249 (1960).

110. ROSEN, L.: Reoviruses in animals other than man. Ann. N.Y. Acad. Sci. **101**, 461—465, Article 2 (1962).

111. ROWE, W. P., R. J. HUEBNER, L. K. GILMORE, R. H. PARROTT, and T. G. WARD: Isolation of a cytopathogenic agent from human adenoids undergoing spontaneous degeneration in tissue culture. Proc. Soc. exp. Biol. (N.Y.) **84**, 570—573 (1953).

112. ROWE, W. P., S. G. BAUM, W. E. PUGH, and M. D. HOGGAN: Studies of adenovirus SV_{40} hybrid viruses. I. Assay system and further evidence for hybridization. J. exp. Med. **122**, 943—954 (1965).

113. RUCKLE, G.: Discussion Part III — Virus in search of disease. Ann. N.Y. Acad. Sci. **67**, 355—362 (1957).

114. RUCKLE, G.: Studies with the monkey-intra-nuclear-inclusion-agent (MINIA) and foamy agent derived from spontaneously degenerating monkey kidney cultures. I. Isolation and tissue culture behavior of the agents and identification of MINIA as closely related to measles virus. Arch. ges. Virusforsch. **8**, 139—167 (1958).

115. RUCKLE, G.: Studies with the monkey-intra-nuclear-inclusion-agent (MINIA) and foamy-agent. II. Immunologic and epidemiologic observation in monkeys in a laboratory colony. Arch. ges. Virusforsch. **8**, 167—182 (1958).

116. RUSTIGIAN, R., P. JOHNSTON, and H. REIHART: Infection of monkey tissue cultures with virus-like agents. Proc. Soc. exp. Biol. (N.Y.) **88**, 8—16 (1955).

117. SABIN, A. B., and A. M. WRIGHT: Acute ascending myelitis following a monkey bite, with the isolation of a virus capable of reproducing the disease. J. exp. Med. **59**, 115—136 (1934).
118. SATTAR, S. A., and K. R. ROZEE: Studies on the biological properties and classification of SV$_4$ virus. Canad. J. Microbiol. **11**, 325—335 (1965).
119. SAUER, R. M., J. E. PRIER, R. S. BUCHANAN, A. A. CREAMER, and H. C. FEGLEY: Studies on a pox disease of monkeys. I. Pathology. Amer. J. vet. Res. **21**, 377—380 (1960).
120. SCHRIER, A. M.: Editor's notes, Primate Newsletter **5**, No. 4, ii (1966).
121. SCHULTZ, E. W., and K. J. HABEL: SA Virus — a new member of the Myxovirus group. J. Immunol. **82**, 274—278 (1959).
122. SHAH, K. V.: Neutralizing antibodies to simian virus 40 (SV$_{40}$) in human sera from India. Proc. Soc. exp. Biol. (N.Y.) **121**, 303—307 (1966).
123. SHEIN, H. M., and J. F. ENDERS: Transformation induced by simian virus 40 in human renal cell cultures. I. Morphology and growth characteristics. Proc. nat. Acad. Sci. (Wash.) **48**, U.S.A., 1164—1357 (1962).
124. SOPER, W. T.: Monkey B virus. A review of the literature. Technical study 17. Biological Warfare Laboratories, Fort Detrick, Maryland (1959).
125. STILES, G. E., J. L. BITTLE, and V. J. CABASSO: Comparison of simian foamy virus strains including a new serological type. Nature (Lond.) **201**, 1350—1351 (1964).
126. SUMNER-SMITH, G.: B virus in association with a monkey colony at a department of psychology. Primate Newsletter **5**, 1—4, No. 1 (1966).
127. SWEET, B. H., and M. R. HILLEMAN: The vacuolating virus, SV$_{40}$. Proc. Soc. exp. Biol. (N.Y.) **105**, 420—427 (1960).
128. TAMM, I., and H. J. EGGERS: Differences in the selective virus inhibition action of 2-(Alpha-Hydroxybenzyl-)Benzimidazole and Guanidine HCl. Virology **18**, 439—447 (1962).
129. TAYLOR, J.: Studies on the purification and properties of two RNA-containing viruses, with particular reference to the fractionation of particles in poliovirus suspensions. Thesis, School of Hygiene, University of Toronto, Ontario (1962).
130. TODARO, G. J., H. GREEN, and M. R. SWIFT: Susceptibility of human diploid fibroblast strains to transformation by SV$_{40}$ virus. Science **153**, 1252—1254 (1966).
131. TRIBE, G. W.: An investigation of the incidence, epidemiology and control of simian virus 5. Brit. J. exp. Path. **47**, 472—479 (1966).
132. TYRRELL, D. A. J., F. E. BUCKLAND, M. C. LANCASTER, and R. C. VALENTINE: Some properties of a strain of SV$_{17}$ virus isolated from an epidemic of conjunctivitis and rhinorrhoea in monkeys *(Erythrocebus patas)*. Brit. J. exp. Path. **41**, 610—616 (1960).
133. VANFRANK, R. M.: Lilly Research Laboratories. Personal communication (1960).
134. WALLIS, C., and J. L. MELNICK: Suppression of adventitious agents in monkey kidney cultures. Texas Rep. Biol. Med. **20**, 465—475 (1962).
135. WENNER, H. A.: Personal communication (1958).
136. WOOD, W., and F. T. SHIMADA: Isolation of strains of virus B from tissue culture of cynomolgus and rhesus kidney. Canad. J. publ. Hlth **45**, 509—558 (1954).
137. YOHN, D. S., J. T. GRACE, and V. A. HAENDIGES: A quantitative cell culture assay for Yaba tumor virus. Nature (Lond.) **202**, 881—883 (1964).
138. YOHN, D. S., V. A. HAENDIGES, and J. T. GRACE: Yaba tumor poxvirus synthesis *in vitro*. I. Cytopathological, Histo-chemical, and Immunofluorescent Studies. J. Bact. **91**, 1977—1985 (1966).
139. YOHN, D. S., V. A. HAENDIGES, and J. T. GRACE: Yaba Tumor Poxvirus synthesis *in vitro*. II. Adsorption, inactivation and assay studies. J. Bact. **91**, 1953—1958 (1966).
140. YOHN, S., V. A. HAENDIGES, and E. DE HARVEN: Yaba tumor poxvirus synthesis *in vitro*. III. Growth kinetics. J. Bact. **91**, 1986—1991 (1966).

141. YOHN, D. S.: Personal communication (1967).
142. ZEITLYONOK, N. A., M. YA. CHUMAKOVA, N. M. RALPH, and LO SIAUW GOEN:
     Distribution area and natural hosts of latent viruses of monkeys. Occurrence of
     simian vacuolating virus (SV_{40}) and Herpesvirus simiae in cynomolgus monkeys
     in Indonesia. Acta virol. **10**, 537—541 (1966).

Rhinoviruses

By

D. A. J. Tyrrell

Clinical Research Centre
Common Cold Research Unit

Salisbury, Wiltshire, England

With 15 Figures

Table of Contents

I. History of Virus Isolations

The aetiology of colds was studied for many years before rhinoviruses were cultivated but recent work indicates that many of the cases of common colds occurring in the general population are due to infection with these organisms (TYRRELL and BYNOE, 1966); it is therefore probable that many of the early efforts to study and cultivate "the common cold virus" were, in fact, experiments on one strain or other of the rhinoviruses. Consequently it may be of interest to start this introduction with a brief history of early experiments on the problem of the cause of the common cold.

The first experiment to show that a cold could be due to a virus was that of KRUSE (1914). He made bacteria-free filtrates of the nasal secretions of cold-sufferers and inoculated these into the noses of 36 healthy volunteers. Fifteen of these volunteers developed colds and he concluded that the disease was due to a virus. However the time was not ripe for the exploitation of this discovery. Techniques for the cultivation of viruses were limited to the inoculation of experimental animals and there was a widespread and firmly-held belief that colds were due to the bacteria which could, of course, often be cultivated from the nasal secretion of patients with the disease (see THOMSON and THOMSON, 1932). However, shortly afterwards, FOSTER (1916) repeated and confirmed the experiments of KRUSE but he also claimed to have cultivated a cold-producing agent anaerobically.

Little real progress was made until some particularly thorough studies were carried out by DOCHEZ and his colleagues at the Rockefeller Institute, New York (DOCHEZ et al., 1938). They made long-term observations on normal subjects and showed that colds might come and go but the bacterial flora of the upper respiratory tract remained largely unchanged (SHIBLEY et al., 1926). They also showed that filtered nasal secretions which contained no ordinary bacteria or "filter-passing anaerobes" usually induced typical colds when inoculated into the upper respiratory tract of chimpanzees and also of volunteers who were similarly inoculated; both the animals and volunteers were kept in strict isolation from other sources of infection. They concluded that colds were due to viruses (DOCHEZ et al., 1930). They also claimed that this virus could be propagated in cultures of chick embryo tissue. However, this last point could not be confirmed by ANDREWES and OAKLEY (1933).

In 1946 the COMMON COLD RESEARCH UNIT was set up at Salisbury, England by the Medical Research Council. There ANDREWES and his colleagues began a long series of studies on the viruses of colds using batches of up to 30 isolated human volunteers every two weeks to test for and sometimes to produce the viruses in which they were interested.

A most valuable part of their early work was to find a reliable and agreeable way of isolating volunteers and assessing their symptoms. The volunteers were put in pairs or threes in separate huts. Their daily food and other requirements were brought to them but they were free to leave the huts to walk or play games provided they did not come closer than 30 feet to other people. They were examined daily for several days during a preliminary period of isolation. They were then inoculated with nasal drops which contained either sterile diluent or test material.

Neither they nor the clinical observer had any means of knowing which volunteers had received which material, and this was not revealed until the end of a period of five further days of isolation and observation when the clinician's conclusion had been recorded. (ANDREWES, 1951; TYRRELL, 1963 b). With this procedure there were very few "false positive" results which may easily occur if rigid quarantine is not enforced or enthusiastic volunteers or staff have a clue that they may have had infectious material. Provided groups of at least six or eight are used this is also a reliable, if cumbersome, way of detecting the presence of a cold-producing agent (RODEN, 1958). ANDREWES and his colleagues then showed that certain observations which had been made by others and which suggested that human common cold virus would cause colds in various sorts of animals could not be confirmed experimentally. They also failed to confirm reports that these viruses would multiply in chick embryos (ANDREWES, 1951, 1953). This period of "clearing the ground" was valuable but not very encouraging, but in 1953 a virus was apparently cultivated successfully (ANDREWES et al., 1953). Roller tube cultures of explants of human embryo lung were prepared and inoculated with washings from a patient with a cold. There was no cytopathic

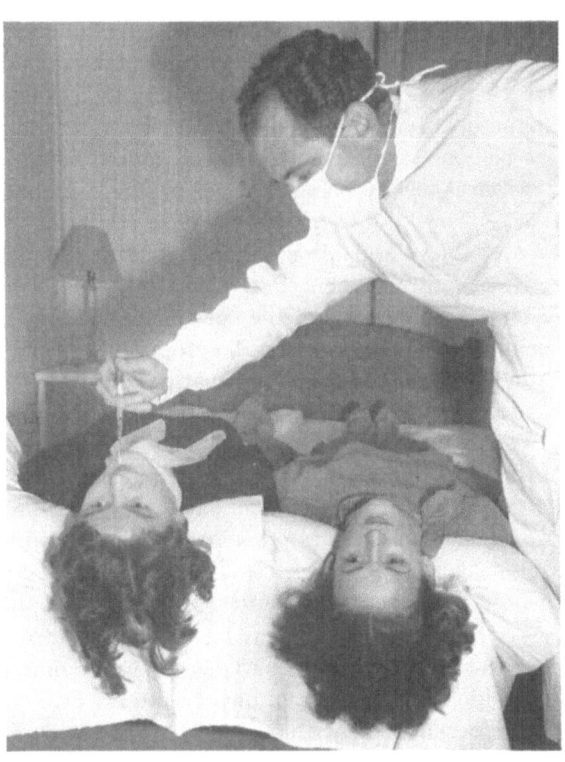

Fig. 1. The technique of inoculating human volunteers in order to induce colds with rhinoviruses and similar agents. The virus is suspended in 1 ml of buffered saline. The volunteer hangs the head back over the end of a bed and keeps it there for 1 minute after the instillation of drops into each nostril. It is forbidden to blow the nose for the next hour. Other workers (KNIGHT, 1964) also spray virus into the pharynx.

effect but the culture fluids were passaged serially. The fluids were also inoculated into volunteers and it appeared that although fluids from uninoculated cultures had no effect, those from inoculated cultures up to the tenth serial passage produced colds. It seems that the virus must have been multiplying, and subsequent studies revealed that it was an H rhinovirus (TYRRELL et al., 1962). At that time however, the results could not be repeated, probably because the cells used in the first experiment were unusually sensitive to the virus and relatively insensitive strains of cells were used in later experiments.

A little later two groups of workers in the U.S.A. recovered a new type of virus which was provisionally designated as ECHO virus type 28. PRICE and colleagues recovered the agent by inoculating monkey kidney cultures with specimens from patients at the Johns Hopkins Hospital, Baltimore (PRICE, 1956; PRICE et al., 1959). They observed a cytopathic effect resembling that due to enteroviruses. They recovered the virus in this way from some patients and detected rising titres of neutralizing antibody. After a time, however, the virus apparently disappeared from the area, but while it was present they accumulated evidence from family studies that the virus was aetiologically related to minor respiratory illness and that persons possessing antibody did not become infected. The illness of the patients was clinically that of a typical common cold, in which nasal discharge was the main symptom. After passage in tissue cultures JH virus was administered to volunteers at Salisbury. It clearly multiplied in the nasopharynx and could be passed from volunteer to volunteer, but it could not be conclusively shown that it caused colds — perhaps because it had lost some virulence by being adapted to monkey kidney cells (TYRRELL and BYNOE, 1958).

MOGABGAB and his colleagues studied naval personnel at a U.S. Navy base at Great Lakes, Illinois (PELON et al., 1957). Their patients also had minor upper respiratory illnesses and the specimens were inoculated into roller tube cultures usually of monkey kidney. A cytopathogenic virus was isolated which was designated 2060. It produced a cytopathic effect which resembled that due to enteroviruses and was found to grow freely under a rather limited range of pH. Rising titres of antibody were detected in infected subjects and the characteristics of the virus like those of JH made it clear that it was different from viruses such as the influenza viruses and adenoviruses which at that time were the only known causes of respiratory disease. These studies did not prove that the virus caused colds but later work by JACKSON et al. (1960a) showed that the virus caused colds in volunteers and confirmed that it was antigenically very similar to JH.

At Salisbury it was found that the type of cultures which would support the growth of ECHO virus type 28 would not reveal the presence of a virus in washings which were known to induce colds in human volunteers. It nevertheless seemed likely that if appropriate cells and conditions could be found these viruses might grow in tissue cultures. However, previous experience suggested that it might be very difficult to forsee or predict what combination of cells and conditions would be required to grow them. It was therefore decided to attempt to grow these "uncultivable" viruses in several cell types, in several media, and at the temperature of the nose, namely 33°C, as well as at the usual temperature of 36°C or 37°C. Roller tube cultures were therefore inoculated with infected nasal washings and the culture media were collected at intervals and stored at −70°C. Later the fluids were thawed and pooled, and the pools were inoculated into volunteers. As expected only limited success was achieved at first, but it was found that fluids from human embryo kidney cultures maintained at 33°C in medium 199 for five days induced colds in volunteers although the cultures appeared normal microscopically and when tested in other ways (TYRRELL et al., 1960). They were tested for virus interference and it was found that at the time when the fluid induce colds in volunteers the cells of the cultures were partially resistant to infection with a number of viruses, of which ECHO virus 11 was the most convenient to use (HITCHCOCK and TYRRELL, 1960). The medium was next modified by the addition of bovine plasma albumin and glucose and the viral interference was then found to be more marked. In addition, using the modified medium it was possible to transmit the interference effect and the capacity to produce colds through eight passages in tissue culture, whereas previously only two passages were successful. Finally, following some accidental changes in the medium it was discovered that a cytopathic effect developed in infected cultures maintained with a rather low concentration of sodium bicarbonate (TYRRELL and PARSONS, 1960). This cytopathic effect resembled that due to enteroviruses, from which however, these Salisbury viruses, as they were then called, were entirely distinct antigenically. It was found that viruses of this type could be recovered from 20—30% of patients with colds (TYRRELL and BYNOE, 1961) but that the viruses were not uniform; most of them multiplied only in human kidney cells, but others grew equally well in monkey kidney cells. The former were therefore called H strains and the latter M strains. There were several different serotypes of both H and M strains although all that were tested produced colds in volunteers (TAYLOR-ROBINSON and TYRRELL, 1962a). The growth requirements and the type of illness they produced indicated that these viruses were so highly adapted to life in the nose that the name rhinovirus was tentatively suggested; although other workers isolated many other strains under somewhat different conditions and gave them different names such as coryzavirus (HAMPARIAN et al., 1961), enteroviruses (JOHNSON and ROSEN, 1963), muriviruses and respiroviruses (MOGABGAB, 1962), ERC group (HAMPARIAN et al., 1963); the name rhinoviruses was agreed upon in the end (ANDREWES, 1961—62; ANDREWES et al., 1961; TYRRELL and CHANOCK, 1963, VIRUS SUBCOMMITTEE, 1963) and has since come into general use.

HAYFLICK and MOORHEAD (1961) developed techniques for the regular serial propagation of strains of human embryo diploid fibroblast cells. These cells were different from malignant or transformed human cell lines which were not only heteroploid but apparently immortal; the diploid cells could not be established in permanent culture but began to degenerate and died out usually between the 30th and 50th passage. It was also shown that certain diploid cell strains were highly susceptible to rhinoviruses and could be used for their isolation and propagation and this discovery accelerated progress in isolating and serotyping these organisms. It was also found that rhinoviruses could be adapted to transformed or malignant human cell lines such as HeLa and KB (TAYLOR-ROBINSON et al., 1963; JOHNSON and ROSEN, 1963) and this has assisted greatly the production of virus for antigens and for purification and similar studies.

By 1962 a great many virus strains had been isolated, particularly in the U.S.A. and several different laboratories were studying their serology. This work began at that time to be co-ordinated under the auspices of the World Health Organization. The object was to obtain well-characterized pure stocks of prototype virus, distribute these to collaborating laboratories, and finally to agree on which viruses were antigenically distinct and to give them serial numbers. The first group of 55 rhinovirus serotypes has now been agreed upon and numbered (KAPIKIAN et al., 1967).

More recently still it has been found possible to cultivate further viruses. These are viruses which cannot apparently be propagated or detected by inoculating specimens into the most carefully prepared and sensitive cultures of human embryo kidney cells or human fibroblast cells. They do however, multiply in the ciliated epithelial cells of human embryo nasal epithelium maintained in organ cultures. Many of these "new" viruses have been shown to be rhinoviruses although several others are not (TYRRELL and BYNOE, 1965, 1966; ALMEIDA and TYRRELL, 1967). The ciliated epithelium of organ cultures is often destroyed by rhinoviruses so that ciliary movements cease and in this way organ cultures may be used to detect rhinoviruses. Some of these rhinoviruses, after a few passages in organ cultures can be grown in tissue cultures of human embryo fibroblasts, while others seem to be unable to reproduce in anything but human ciliated respiratory epithelial cells (HOORN and TYRRELL, 1966; HIGGINS, 1967).

II. Classification and Nomenclature

Soon after the first group of rhinoviruses had been studied it was clear that the rhinovirus particles were ether-stable and of the same size and density as well-studied enteroviruses such as the polioviruses. It was suspected and later confirmed that they also contain infectious RNA. Furthermore the cytopathic effect produced was similar qualitatively to that of enteroviruses although the range of susceptible cells and the optimal conditions for growth were different. At the same time it was recognised that in man the rhinoviruses seemed to be particularly well adapted to multiply in the nose and to cause nasal and other respiratory symptoms and also that they were almost never found in the faeces; by contrast, the enteroviruses multiplied freely in the lower alimentary tract and were not regularly associated with upper respiratory disease. The optimal growth requirements, such as a temperature of 33°C rather than 37°C did not unequivocally distinguish rhinoviruses from enteroviruses, but rhinoviruses were found to be inactivated rapidly in weakly acidic solutions (pH 3—5) which had no effect on enteroviruses. These facts, which are elaborated in later sections, are the basis on which the viruses are now grouped together and called rhinoviruses. However, this grouping is not without difficulties, some of which should be discussed here.

It is generally agreed that the classification of viruses by the type of disease they produce has little or nothing to commend it. It is also clear that rhinoviruses are of the same size, shape and basic structure and have the same type of nucleic acid and presumably the same method of reproduction as enteroviruses such as polioviruses, Coxsackie viruses and ECHO viruses. There is therefore no doubt that rhinoviruses and enteroviruses all belong to one large group of viruses. It would be straining the term to enlarge the group of enteroviruses to include agents which rarely, if ever, grow in the lower alimentary tract; therefore a new group name was proposed, namely that of picornaviruses (INTERNATIONAL ENTEROVIRUS STUDY GROUP, 1963) to include all those small (*pico*) viruses containing infectious ribonucleic acid (RNA). The properties of typical members of the group are shown in Table 1.

The subdivision of the viruses shown is generally useful and the viruses which have the *in vitro* properties of rhinoviruses generally seem to produce rather

Table 1. *The Properties of Some Typical Picornaviruses*

	Enteroviruses			Rhinovirus
	Poliovirus	Coxsackie virus	ECHO virus	M and H strains
Particle: Symmetry Nucleic acid	cubical RNA	cubical RNA	cubical ? RNA	cubical ? RNA
Stability to ether and chloroform	stable	stable	stable	stable
Stability at pH 3—5	stable	stable	stable	unstable
Main site of infection in man	upper and lower alimentary tract + C.N.S.	upper and lower alimentary tract + C.N.S., respiratory tract and muscle especially in Group B	upper and lower alimentary tract + C.N.S., skin etc.	upper and sometimes lower respiratory tract
Host range of virus Monkey C.N.S.	often paralysis	rarely paralysis	rare	?
Suckling mouse	None	Group A — myositis Group B — encephalitis, pancreatitis etc. Sometimes little or none	none	none
Growth in tissue culture	grow well in primary and secondary monkey and human kidney and fibroblasts and transformed cell lines.	Certain A strains grow in human amnion and fibroblasts and one or two in monkey kidney cells. The B strains resemble polioviruses.	grow freely in monkey kidney and primary human kidney. Often grow poorly in transformed cells.	grow only in human embryo kidney or fibroblasts (the latter after passage in tissue culture). Can be adapted to certain transformed cell lines.

similar mild infections of the upper respiratory tract. On the other hand, some of the subdivisions within the rhinovirus and enterovirus groups are based largely on the host range of viruses and these are not clearcut in every case. For example, Coxsackie viruses of group A are generally speaking enteroviruses which produce a necrosis of the striated muscle of suckling mice. But in the case of Coxsackie virus A23, which is also called ECHO virus 9, some strains may be isolated and propagated regularly in suckling mice, others may be isolated in tissue culture and then adapted to mice and still others can be grown only in tissue cultures like a typical ECHO virus (e.g. TYRRELL et al., 1958). It is therefore not surprising to find viruses which have biological properties which put them on the borderline between rhinoviruses and enteroviruses. For example, Coxsackie virus A24 which was first recovered in suckling mice is now commonly recovered from the faeces of children by inoculation into human embryo kidney or amnion cells and will not multiply in suckling mice (HENIGST, 1965; BEH-BEHANI, 1966). It therefore has an *in vitro* host range reminiscent of a rhinovirus although it grows freely in the alimentary tract and apparently does not cause respiratory disease when inoculated into the upper respiratory tract of volunteers (KASEL and KNIGHT, 1963). Coxsackie virus A21 is isolated most frequently from respiratory secretions obtained from patients with acute upper respiratory infections. However, the prototype virus was obtained in suckling mice from the faeces of a child and a virus isolated in HeLa cells has been adapted with difficulty to grow in suckling mice. Nevertheless recent work shows that the virus may be most readily isolated and propagated in tissue cultures of human kidney or human embryo fibroblast cultures (e.g. JOHNSON et al., 1962; BUCK-LAND et al., 1965) which also support the growth of H rhinoviruses, although in the case of Coxsackie virus A21 the optimal temperature of growth is above 33°C. Both these Coxsackie viruses are apparently typical enteroviruses when studied in the laboratory, and in particular both are acid-stable like all other known enteroviruses (KETLER et al., 1962; KAWANA et al., 1965). Similarly, Coxsackie viruses of group B are now isolated by inoculation of tissue cultures and are recovered as often from the respiratory secretion of cases of febrile respiratory infections as they are from the alimentary tract of cases of pleuro-dynia or aseptic meningitis with which they were first associated (e.g. WORKING PARTY, 1965; HOLZEL et al., 1965a).

In the first experiments at Salisbury it was noticed that within the rhino-virus group there were strains which would multiply freely within monkey kidney cells and others which would grow only in human cells — either human embryo kidney or human embryo fibroblast cells. This led to the separation of M (monkey) and H (human) strains and subsequent experience has shown that the M character is possessed by a rather limited number of widely distributed serotypes, while strains with the H character belong to many more serotypes. It is also clear that the M character is not completely clear cut, for some viruses, for example the 30/60 strain may be grown in monkey kidney cells only with considerable difficulty. In addition high concentrations of certain M and H strains will produce a cytopathic effect in another type of non-human cells, namely calf kidney cells, although these viruses have not been successfully adapted to this cell type (VARGOSKO, unpublished). Similarly, after adaptation to human fibroblasts high concentrations of H rhinoviruses will produce a typical cytopathic effect when inoculated into monkey kidney cultures (DOUGLAS et al., 1966b).

We have recently been tempted to use the term O-strains for viruses which are cultivated in organ cultures of human respiratory epithelium. But this is an over-simplification for although there are some strains which multiply only in organ cultures and not in tissue cultures of human and other cells, there are other

Table 2
A. Certain General Properties of Typical Rhinoviruses

	M	H	O–H	O
Passage through 50 mμ filter	+	+	+	+
Stability to ether or chloroform	+	+	+	+
pH 3–5	0	0	0	0
Inhibition by BUDR	+	+	+	+
Growth and cytopathic effect in:				
1. Monkey kidney	+	0	0	0
2. Human embryo kidney, epithelium or lung fibroblast cells	+	+	+[1]	0
3. Human embryo nasal or tracheal cells in organ culture	+	+	+	+

[1] after passage.

B. Physical Properties of Certain Rhinoviruses

	1A	1B	2
Size by filtration (mμ)		31	31
by electron microscopy (mμ)		22–23	30
Sedimentation coefficient	equal to poliovirus	165S	155S
Density in CsCl$_2$ of upper component	1.30	1.28–1.30	1.31–1.30
Density in CsCl$_2$ of lower component	1.41	1.38–1.39	1.40–1.41

DIMMOCK and TYRRELL (1964). CHAPPLE and HARRIS (1966). MCGREGOR et al. (1966). DANS et al. (1966).

viruses which can at present be recovered from clinical material only by inoculation of organ cultures but can later be adapted to grow in human embryo fibroblasts. The former might be called O and the latter O-H strains, but all of them so tested have the same physical properties and the same optimal temperature and pH for growth as typical M and H rhinoviruses (Table 2).

In conclusion it seems reasonable to separate a group of acid-labile human picornaviruses and to name them rhinoviruses since their most striking characteristic from the point of view of their ecology and pathogenesis seems to be their ability to infect the human nose. Naturally a number of other related and unrelated viruses also infect the human nose. There are some viruses in fact which represent intermediate organisms between typical enteroviruses and rhinoviruses. Within the rhinovirus group there are extremely fastidious organisms which multiply only in human ciliated epithelial cells, and other strains which infect and damage human fibroblasts and embryo kidney cells and sometimes monkey kidney cells. All so far studied seem able to produce colds in man. It has been pointed out elsewhere that Foot-and-Mouth disease virus is an acid-labile picornavirus of cattle and might be considered to be a rhinovirus if acid lability were to be a sole criterion (PLUMMER, 1965). However, it causes a disease quite different from those produced by the rhinoviruses of cattle, horses and cats which have so far been described and it seems premature at the moment to force it into the same group as these. The work of numerical taxonomists indicates that an objective method of classification would be based on a large number of independent characters of the organism. The trouble at the moment is that we possess too few such characters to use this approach for the separation of viruses within the picornavirus group. Some hold the view that because picornaviruses cannot be separated into completely distinct subgroups they should be numbered in serial sequence according to the date on which they were discovered or certain tests were completed (ROSEN, 1965). The present author has no sympathy with this view. Because of the large number of organisms involved it is better to place them in groups, and so to gather together viruses with broadly similar characteristics. By so doing one need not blind oneself to the fact that such a grouping inevitably represents an oversimplification of the subtle gradations which exist between different varieties of even such a simple organism as a rhinovirus. ANDREWES has picturesquely called this the "viro-astronomical" concept, likening groups of viruses to constellations. He feels that it is better to be able to place the majority of viruses clearly in one group or another than to abandon attempts at classification because a few viruses are intermediate and have to be allocated to a group in a rather arbitrary way.

III. Properties of the Virus
A. Antigenic Structure

Early experiments on the antigenic composition of rhinoviruses were hampered by difficulties in performing neutralization tests, because the sensitivity of successive batches of monkey-kidney or of human embryo-kidney cells varied considerably and the neutralizing titre of serum depended to a great extent on the amount of virus used in the test. The problem was solved in two ways. Firstly, it was found that microplaques were formed in roller tube cultures and that, provided these were counted before about the third day of incubation, the counts were directly proportional to the concentration of virus added (PARSONS and TYRRELL, 1961). It was then shown that for several hours after the mixture of antiserum and a rhinovirus the inactivation of infectivity as detected by microplaque counts was described by a simple exponential curve and that the rate of inactivation was directly proportional to the amount of serum added (TAYLOR-ROBINSON and TYRRELL, 1962b). It was thus possible to use the reduction in microplaque count as a delicate and precise method of measuring antiviral activity, for a reduction of 50% or more in plaque count was significant even when quite small numbers were counted; since the result depended on the ratio between the number of plaques produced by the virus with and without the admixture of serum the result was largely independent of variations in the exact sensitivity of the cells. The second solution to the problem was to use cultures of human diploid fibroblast strains as the test object. These varied less in their sensitivity to the virus than did the cells previously used and it was therefore possible to perform the usual type of end-point neutralization tests using serial dilutions of serum and 10—100 infectious doses of virus (e.g. KETLER et al., 1962; JOHNSON and ROSEN, 1963). This test was slightly less sensitive than the microplaque reduction test but detected almost all the neutralizing antibody rises which followed natural infections and was satisfactory for serotyping when using specific animal sera (TAYLOR-ROBINSON et al., 1963). The test requires less time for examination of the cultures than the microplaque reduction test. It has therefore been widely used, particularly where large numbers of tests have to be performed and there is an abundance of cultures. HAMPARIAN and co-workers (1964) adapted their viruses to a sensitive line of HeLa cells and then did end-point neutralization tests in tube cultures of these. The sensitivity of HeLa cells to rhinoviruses can be enhanced still further by addition of magnesium ions and this has made it possible to cary out plaque reduction tests (FIALA and KENNY, 1966) and colour tests (STOTT and TYRRELL, 1967) similar to those used in work with certain enteroviruses. Two other methods, namely reduction of plaques in monolayers of human diploid cells under agar (PORTERFIELD, 1962) and a microtest using these cells and observing them under the microscope in disposable plastic trays (GWALTNEY, 1966) have been described but have proved difficult to reproduce regularly in other laboratories.

When first isolated, ECHO 28 virus was shown to be distinct from all the then known enteroviruses (PELON, 1961). Preliminary studies of the "Salisbury strain" rhinoviruses showed that they represented six serotypes, one of which, a M strain, was related to the ECHO 28 virus and all of which were unrelated to the then known enteroviruses (TAYLOR-ROBINSON and TYRRELL, 1962a). The viruses used for these tests had not been "purified" by limit dilution or plaque-picking procedures and the tests were performed by microplaque neutralization tests with rabbit sera. The coryza viruses studied in Philadelphia seemed to be antigenically distinct from these, but again were tested without "purification" and using low titre guinea pig sera (KETLER et al., 1962). Strains studied in Bethesda (JOHNSON and ROSEN, 1963) and Chicago (e.g. HAMRE et al., 1964) were tested after limit-dilution-purification and using high-titred sera and appeared to be different from the Salisbury strains. It was therefore thought desirable that all the apparently new serotypes should be tested against each other using a uniform technique; in all cases, "purified" viruses should be employed and sera should be prepared in large batches. A study was therefore set up under the auspices of the WHO as a collaborative venture which was to be co-ordinated by the International Reference Centres for Respiratory Virus Diseases in Salisbury, England, and in Bethesda, U.S.A. Most of the laboratories were supported by the Vaccine Development Board of the U.S.

National Institutes of Health. This Board also backed the production of anti-sera and the setting up of a rhinovirus reference laboratory which was to dupli-cate and check the neutralization tests performed by each laboratory which contributed to the study one or more new viruses and the corresponding anti-sera. It was hoped that out of this co-operative venture would emerge carefully checked laboratory results which would warrant the allocation of viruses to a number of agreed serotypes, and also a supply of exhaustively tested sera and virus seeds which could be used for reference purposes and for the testing of further rhinoviruses isolated.

The chart given as Table 12 is a summary of the results collected and also gives the names of many of the laboratory workers who contributed their results. Several points of importance emerge.

Firstly there are a great many distinct serotypes of rhinoviruses. Although the titres of many of the sera were high, several thousand in fact, there were many occasions on which no neutralization was detected with serum diluted only 1/20 and tested with scores of other viruses. Although not shown in this table the viruses had also been tested against high titre antisera prepared against prototype enteroviruses. Slight cross reactions have been detected, but these have been very rare. On only one occasion was there a substantial degree of cross reaction between two rhinoviruses. This was the two-way cross reaction between ECHO 28 and the Salisbury B 632 strain. So far "broad" and "narrow" reacting strains of viruses, resembling those found earlier among the enteroviruses, have not been found.

Secondly, occasional minor cross reactions do occur. It is not possible to establish their significance completely at the moment. Usually the pre-immuniza-tion serum specimen from the animal concerned contained no antibodies to the virus in question. Often however it was found possible to eliminate these cross reactions by adsorbing the sera with preparations of human cells or human liver powder. Cross reactions have been found particularly with sera from calves (FENTERS et al., 1966), while sera from guinea pigs and goats seem to give more specific results. As tissue cultures containing rhinoviruses are poor antigens for animals they were mixed with FREUND's adjuvant and administered several times and were sometimes purified and concentrated by ultracentrifugation and treatment with fluorocarbon before this. It is therefore more impressive that the sera obtained were so specific rather than that they cross reacted[1].

It has been shown that following natural infections antibody responses may be detected against a number of different picornaviruses, particularly, when complement-fixation tests are used. It had of course been known for some years that in natural infections with enteroviruses one might detect antibody responses against several different serotypes by complement fixation and also by neutraliza-tion tests. The latter probably often represent "recall" or "boost" responses in subjects who have previously been infected with several related viruses; this

---

[1] It is satisfactory that a clear cut numbering system has been agreed on but it is necessary to warn the reader that a numbering system devised by the Merck group for coryzaviruses, was then called by them a rhinovirus numbering system. They have reported viruses using this system which do not agree with the international system and should not be used (HAMPARIAN et al., 1964a, b).

is particularly strongly suggested by experimental infections with a series of entero-viruses. However, the antibody responses of one group of 18 child and adult patients from whom 12 different rhinoviruses were recovered showed small rises in the titre of homotypic antibody and only 3 rises in one patient against 11 other rhinoviruses (HAMPARIAN et al., 1964a); in addition volunteers were infected in isolation with ECHO virus 11, Coxsackie virus A21 or one of two rhinoviruses, and there was no evidence of heterotypic responses by neutralization and similar tests (BUCKLAND et al., 1964). Finally rabbits which were repeatedly inoculated with three different rhinoviruses in succession showed no evidence of heterotypic responses against any of the three used in the test (SCHILD, personal communication). On the other hand when calves were immunized by intramuscular injection of successive different human rhinoviruses, there appears to be booster responses in the titres against first viruses given when later, apparently quite distinct antigenic types were given (FENTERS et al., 1966)[2]. It was suggested at first that there was so much cross reaction among rhinoviruses that a few diagnostic antigens would suffice to recognise infection with most members of this group and that a vaccine containing a few components would protect against infection with many of them (MOGABGAB, 1963b). Recent experience does not support this view — in particular the protection afforded by vaccination seems to be effective only against infection with the serotype used in the vaccine (SCIENTIFIC COMMITTEE, 1965; MUFSON et al., 1963), although only a few sero-types have been tested so far.

Concentrated preparations of rhinoviruses will fix complement, apparently specifically, with most human sera. Using type 2 (HGP) virus most sera are positive from about one year of age up to 70 or 80 years. Many young subjects have complement-fixing antibody and no neutralizing antibody and unchanging titres of CF antibody are observed in subjects infected with the virus or success-fully vaccinated with it (CHAPPLE et al., 1967). It therefore seems that in addition to a strain-specific antigen the HGP rhinovirus produces a broadly reacting antigen. Antibody to the latter is probably induced in the first place by infection with an agent other than the HGP strain, probably another rhinovirus, and this antibody persists at high titre through life. Recent work suggests that this group-specific type of antibody response is shown by less dense virus particles, comparable to the C particles of poliovirus type 1 and to similar antigens found in preparations of Coxsackie viruses (DANS et al., 1966).

It may be possible in the future to prepare suspensions of full virus particles which react strain specifically in the complement-fixation test, but this is not so at the moment, and so we still lack an *in vitro* test for the specific antigens and antibodies of these viruses. Specific complement-fixation, immuno-diffusion or other type of test may be developed soon. Preliminary results indicate that rhinoviruses inhibit the haemagglutination of trypsin-treated red cells suspended in buffered glucose. This inhibition can itself be inhibited by antisera and this procedure may some day be made the basis of a useful *in vitro* serological test (TYRRELL et al., 1967).

---

[2] These results may indicate that there are subgroups of rhinoviruses which have some distant antigenic relationships with each other.

B. Resistance to Physical and Chemical Reagents

Experience over many years with volunteers has shown that if nasal secretion or nasal washings are mixed with a final concentration of 50% bacteriological broth as is usually done at Salisbury or with 2.5% haemoglobin (JACKSON, 1958) the infectivity of "cold viruses" for volunteers will be preserved for many years when frozen at about −70°C. In some recent studies rhinoviruses in clinical specimens have been satisfactorily preserved in transport media containing 2% bovine plasma albumin or veal infusion broth at −70°C (WORKING PARTY, 1965; HAMPARIAN et al., 1961). Infectivity may be preserved for a while at −20°C but seems in some cases to be slowly lost at −40°C (BLOOM et al., 1963).

The infectivity of tissue culture fluids is slowly lost at 4°C and rapidly at 37°C or higher and the inactivation process has been studied in detail by DIM-MOCK (1967). It appears that different processes determine inactivation of HGP virus above and below about 37°C. The process occurring at the higher temperature has a high activation energy of about 101 Kals/mole which falls to 75 Kals/mole in the presence of molar magnesium chloride. That occurring at the lower temperature has a lower activation energy, 19 Kals/mole, very close to that of the loss of infectivity of isolated poliovirus RNA and is not reduced by Mg. At the higher temperature antigenic activity is impaired by heating while the RNA within the particle apparently retains its infectivity, whereas at lower temperature the antigenicity is unaffected and the infectivity of the extractable RNA is lost (DIMMOCK, 1967). The results are similar to those obtained with the LSc strain of type 1 poliovirus except that this rhinovirus, like many others, is relatively more stable in the high temperature range and that M MgCl$_2$, which completely stabilizes all enteroviruses at 55°C, is much less effective in stabilizing rhinoviruses at this temperature (DIMMOCK and TYRRELL, 1962, 1964). These facts suggest that there is a difference between the stability of the proteins of rhinoviruses and of enteroviruses and this is confirmed by the further fact that rhinoviruses are acid-labile and rapidly lose infectivity between pH 3 and 5 while enteroviruses are stable down to pH 2 (DIMMOCK and TYRRELL, 1962, 1964; KETLER et al., 1962), and rhinoviruses are inactivated by urea while polioviruses are generally not (DIMMOCK, 1968). The relative stability of rhinoviruses at high temperature in neutral solutions and their instability at lower temperatures in acid solutions have been confirmed for a large number of serotypes and the acid lability test is now widely used to distinguish between rhinoviruses and enteroviruses when they are isolated from clinical specimens (see TYRRELL and CHANOCK, 1963, and most papers describing the isolation of rhinoviruses since that date).

All rhinoviruses tested are unaffected by treatment with ether, chloroform or fluorocarbon (TYRRELL and PARSONS, 1960; PELON, 1961; KETLER et al., 1962; GWALTNEY and JORDAN, 1964 and others). Freezing and thawing has little effect on infectivity (PARSONS and TYRRELL, 1962). However caesium chloride in high concentrations and vigorous resuspension from a pellet obtained in an ultracentrifuge may both reduce infectivity (CHAPPLE and HARRIS, 1966; McGREGOR et al., 1966; DANS et al., 1966). When allowed to dry in air on a glass surface or in airborne droplets much of the infectivity is lost in a few minutes

(BUCKLAND and TYRRELL, 1962). However, freeze-drying is not so deleterious and when lyophilized in a final concentration of 5% dextran and 5% glucose there may be no detectable loss of virus infectivity by M strains and not more than 10-fold by H strains; the residual infectivity may be satisfactorily preserved for years at 4°C (TYRRELL and RIDGEWELL, 1965).

C. Physicochemical Structure

Because rhinoviruses do not usually grow to high titre in tissue cultures, particularly when they have only recently been isolated, early attempts to obtain preparations for electron microscopy failed or were only partly successful. Recently it has been found that starting material with a high titre may be obtained by inoculating large doses of a well-adapted strain of virus to cultures of susceptible lines of human heteroploid cells such as KB and HeLa. The virus produced has been concentrated and purified in various ways. CHAPPLE adsorbed it onto aluminium phosphate gel, eluted it with phosphate buffer treated with fluoro-carbon and fractionated it on a caesium chloride density gradient in the ultra-centrifuge (CHAPPLE and HARRIS, 1966). DANS et al. (1966) processed similar tissue culture fluids and McGREGOR et al. (1966) either deposited the virus onto a "cushion" of caesium chloride solution to avoid damage or deposited the virus in the usual way and allowed the virus pellet to stand in fluid overnight before resuspending it.

D. Electron Microscopy

Electron microscopy of rhinovirus type 2 (HGP strain) shows numerous outlines of uniformly sized particles about 30 mμ in diameter (CHAPPLE and HARRIS, 1966). The outlines are often hexagonal suggesting a symmetrical type of structure and dis-integrating particles which are occasionally seen suggest that there is an outer "shell" containing subunits (Fig. 2b). Particles may be seen which are of the same size and apparently "hollow" or "empty" when examined in phosphotungstate. Particles which are presumed to lack nucleic acid because they are found in the less dense fraction on a caesium chloride gradient may also show some "full" and many "empty" forms, and similarly dense particles may appear empty, so it is not wise to interpret the electron micrograph appearance of this particular virus as meaning the presence or absence of nucleic acid (CHAPPLE and HARRIS, 1966). The subunit structure is difficult to resolve but certainly seems to resemble that of poliovirus although it gives the impression of being rather more loosely knit. MAYOR (1964) believes that she has evidence that there is icosahedral symmetry based on the rhombic triacontahedron but the data to support this have not yet been presented in full. Similar regular particles were seen in preparations of type 1B and 45 by McGREGOR et al. (1966). They estimate the size as 20—23 mμ. The figure of 17—18 mμ given by HAMPARIAN et al. (1961) was based on particles seen in sections of infected cells and is not acceptable.

E. Filtration and Ultracentrifugation

Before satisfactory electron micrographs had been taken the size of the infectious units of rhinoviruses had been estimated by biophysical methods.

It was found that the retention of several rhinoviruses in filtration through a series of collodion membranes was identical with that of poliovirus and calculation from the average pore diameter of the filters which retained the particles indicated a diameter of 31 mµ (DIMMOCK and TYRRELL, 1964). All the 55 serotypes have been shown to pass a 50 mµ collodion membrane (KAPIKIAN et al., 1967). In these studies virus particles were also sedimented in a low concentration of sucrose and this showed that the sedimentation coefficient of many strains was the same as that of poliovirus. From these data it could be calculated that the density in dilute sucrose was about 1.3 and equal to that of poliovirus. On the other hand several laboratories have shown that on equilibration in a caesium chloride density gradient the density of infectious particles is about 1.4 while it is the less dense non-infectious particle and much of the complement fixing

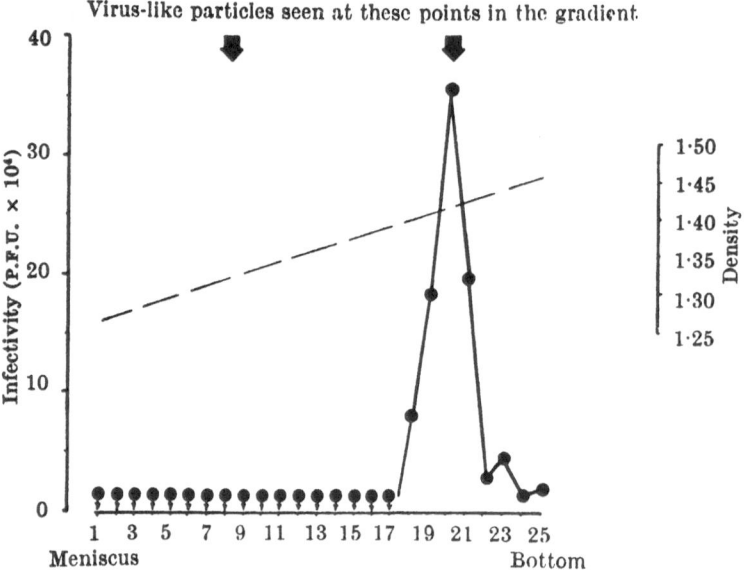

Fig. 2a. The banding of rhinovirus type 2 (strain HgP) by equilibrium centrifugation in a caesium chloride density gradient. Particles of two densities are separated. The more dense is infectious.

activity which are found at about 1.3 (Fig. 2a). The serotypes shown to behave in this way are 1A, 1B and 2 (Table 2B, CHAPPLE and HARRIS, 1966; DANS et al., 1966; McGREGOR et al., 1966). This suggests that exposure to caesium chloride produces some change in the rhinovirus particle which it does not produce in poliovirus particles. Perhaps caesium ions penetrate the particle or dehydrate it. On the other hand, the existence of particles of two densities resembles closely the phenomenon observed in polio- and other enteroviruses, and there is even preliminary evidence that, like enteroviruses, the less dense fraction of rhinoviruses is also less specific antigenically than the more dense fraction (DANS et al., 1966).

It has been shown by all those groups that have studied the matter that the growth of rhinoviruses is not prevented by the presence in the tissue culture medium of 5-fluoro-2'-deoxyuridine (FUDR) or of the corresponding bromo-compound (BUDR). (KETLER et al., 1962; MUFSON et al., 1965; HOORN and TYRRELL, 1966; GWALTNEY and JORDAN, 1964; JOHNSON and ROSEN, 1963; CONNELLY and HAMRE, 1964.) The experiments were well controlled and since these compounds are potent inhibitors of the multiplication of DNA viruses it has been concluded that all rhinoviruses are probably RNA viruses. There has also been a short undocumented report of the extraction of infectious RNA from an H rhinovirus (KETLER et al., 1962). Recently a phenol extract of a concentrated preparation of an M rhinovirus was shown to contain material which was only infectious for cells which were treated with concentrated salt solution and which was susceptible to RNase, insusceptible to antibody and of a low sedimentation coefficient and therefore presumably infectious viral RNA (DIMMOCK, 1966). This direct evidence makes it certain that these H and M rhinoviruses

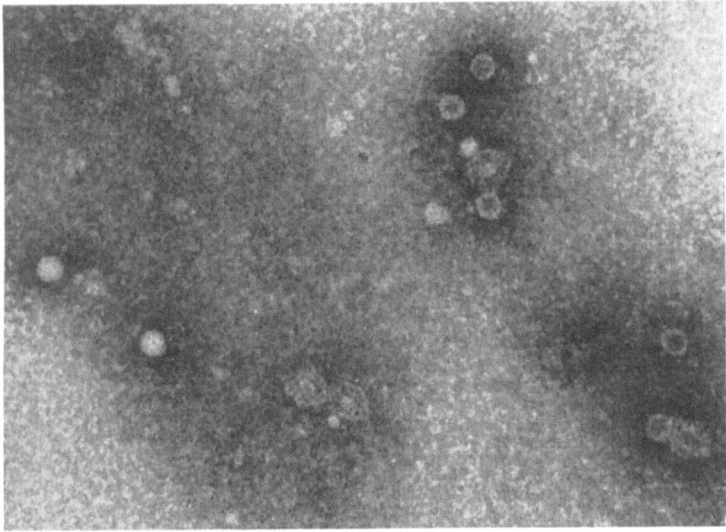

Fig. 2 b. The morphological appearances of a typical "full", "empty" and "unfolding" particle obtained after such a separation and stained by phosphotungstate are also shown 200,000 × (CHAPPLE and HARRIS, 1966). By permission of Nature. (British Crown Copyright reserved. Reproduced with the permission of the Controller, Her Britain Majesty's Stationary Office.)

contain infectious RNA and it is reasonable to assume that it is a character of the whole group.

So far no work on the chemical analysis of the nucleic acid and protein of rhinoviruses has been reported but it should be possible before long to prepare sufficient quantities of sufficiently pure virus to make studies of this kind possible. Meantime it is clear that on morphological and physicochemical examination rhinoviruses are similar to but distinct from the enteroviruses. Although its mode of action is uncertain it may be mentioned at this point that the multiplication of rhinoviruses is usually not inhibited by HBB (2-α-hydroxybenzyl-benzimidazole) (EGGERS and TAMM, 1962) a substance which inhibits the growth of enteroviruses other than Coxsackie viruses. However, some rhinoviruses are inhibited, so that, like heat stability, the test is of little value for classification (HAMRE et al., 1964).

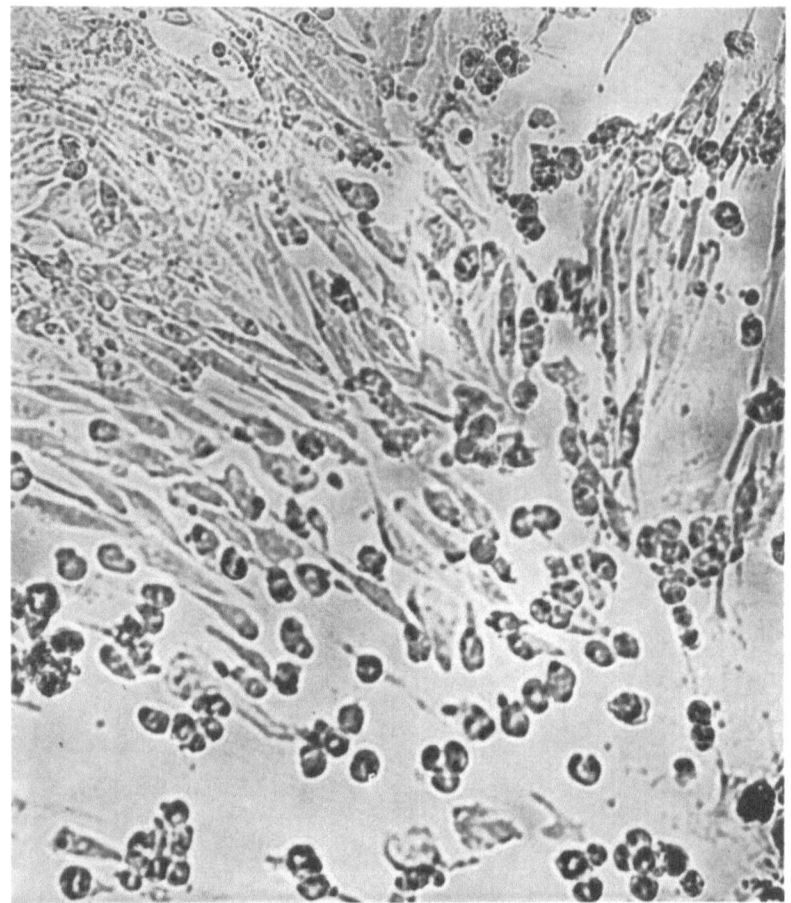

Fig. 3 a – d. The cytopathic effects of rhinoviruses in tissue culture cells. a) Unstained cultures of monkey kidney cells infected with rhinovirus type 2 (By permission Lancet). b) Unstained cultures of human embryo kidney infected with a recently isolated virus (Dr. A. HORNSLETH). c) Unstained cultures of HeLa cells both uninfected and infected with rhinovirus type 2 (E. J. STOTT). d) Unstained cultures of human embryo fibroblasts (strain WI-38) infected 48 hours previously with rhinovirus type 50, 150 × (CONANT and HAMPARIAN). Note the focal degeneration in a – d and the rounding and shrinking of infected cells.

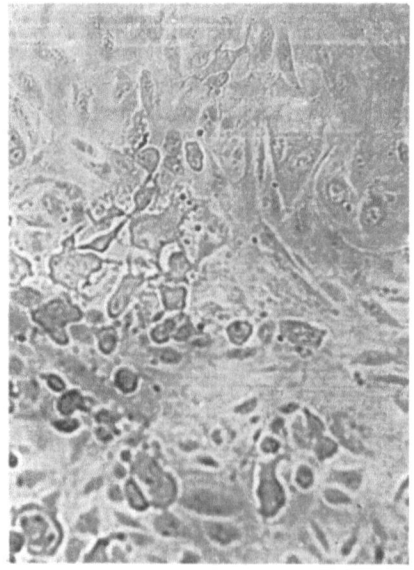

Fig. 3 b

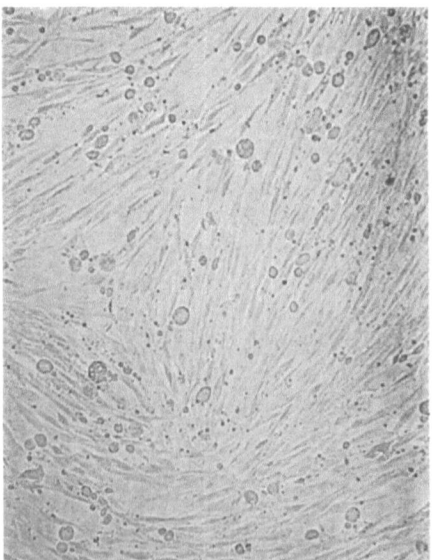

Fig. 3 d

F. Cultivation

The DC strain of rhino-
virus was shown to be growing in cultures of human
embryo lung in roller tubes
at 37°C because although
the cells were unchanged
morphologically the medi-
um produced colds in vol-
unteers (ANDREWES et al.,
1953). The ECHO 28 virus
was first recognised by the
cytopathic effect which it
produced when inoculated
into cultures of rhesus monkey kidney cells in roller
tubes at 35°C (PELON et al.,
1957; PRICE, 1956). It was
later shown that the optimal conditions for the
growth of many rhinoviruses
were the use of trypsin-dis-
persed human-embryo-kid-
ney cells in rolled tubes,
at 33°C using a medium
with a pH between about
6.8 and 7.3 (PARSONS and
TYRRELL, 1960). In such
conditions a cytopathic effect appears in from 18
hours to two weeks depend-
ing on the dose and type
of virus used (Fig. 3). When
high concentrations are used
the cytopathic effect is diffuse but in low concentrations small foci are seen
which may enlarge and coa-
lesce with other foci which
appear later, or, if conditions are unfavourable, may
disappear again. These foci
or microplaques are apparently each initiated by a
single infectious virus particle (PARSONS and TYR-
RELL, 1961). They may be

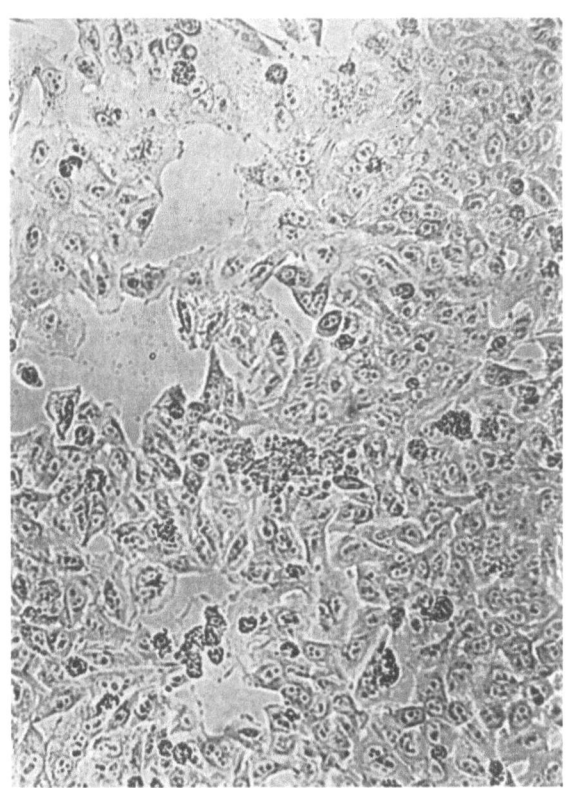

Fig. 3 c

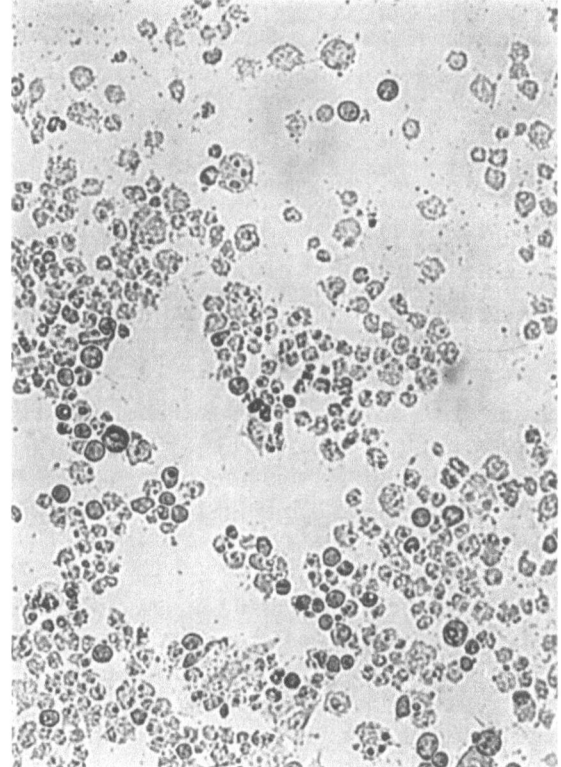

counted under the low power of the microscope, and provided the counts are
made before secondary plaques appear they are an accurate measure of the
amount of infectious virus added to a tube.

Later studies have shown that rhinoviruses may also be grown in human-
embryo-fibroblast-cell-strains in semi-continuous culture (HAYFLICK and
MOORHEAD, 1961; HAMPARIAN et al., 1961; BROWN, et al., 1963). Primary cultures of human-embryo-lung are rather resistant to virus, but after a few serial
passages their sensitivity to virus increases (BROWN and TYRRELL, 1964). However, the final level of sensitivity attained varies considerably from one strain
to another and only a small proportion of embryos (2 out of 33 in the author's laboratory) yield cell strains which are as sensitive as the widely used strains WI-26 and

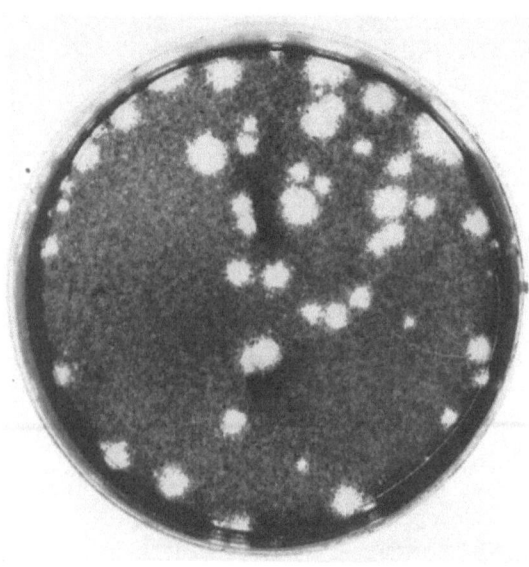

WI-38. Similarly, kidneys
from different embryos yield
primary cell cultures of differing sensitivity to rhinoviruses, and there is a little
evidence that the sensitivity
of the lung and kidney cells
may be correlated (BROWN
and TYRRELL, 1964). Nevertheless it seems that the maximum number of virus iso-
lations are made when both
human embryo kidney and
one or perhaps several lines
of fibroblasts are used (PHIL-
LIPS et al., 1965 b and personal communication; HIG-
GINS, 1966 a; STOTT and
WALKER, 1967). Even if a
cell strain is basically sensitive it may fail to detect
small amounts of virus if the
culture medium is not quite
satisfactory or if the cells

Fig. 4. Plaques produced by rhinovirus type 2 in monolayers
of the M strain of HeLa cells in the presence of 30 mM magnesium chloride under an overlay of NaCl agar. The culture
was fixed with formalin, the agar was removed and the cell
sheets were stained with gentian violet (Preparation by STOTT).

are approaching the end of their capacity to multiply *in vitro*. Recent
studies have suggested that a cell strain derived from human aorta may
be particularly sensitive to rhinoviruses but this has not yet been confirmed in
other laboratories (BEHBEHANI et al., 1965). Cell strains have also been derived
from human embryo kidneys (STOTT and WALKER, 1967) and found to be particularly sensitive to certain strains of rhinoviruses but not better than lung
fibroblasts for others.

Many M rhinoviruses may be quite readily adapted to grow in human heteroploid cell lines such as HeLa, HEp-2 and KB. Not all lines of the two former
cells are equally sensitive (TAYLOR-ROBINSON et al., 1963 a). while the KB cells
seem capable of producing particularly high titres of virus (MUFSON, personal
communication; CHAPPLE and HARRIS, 1966). These cells are not suitable for

general use in titrations as the cytopathic effect is often indistinct, and recently isolated viruses grow poorly.

Initial attempts to produce typical plaques in cells under overlays were unsuccessful (PELON, 1961; PARSONS and TYRRELL, 1961). This was then achieved using HGP, B632 and strain Norman (types 2, 1b and 5) viruses in human embryo fibroblast cells under an agar overlay (PORTERFIELD, 1962). Others have improved the results by adding dextran sulphate or DEAE dextran to the standard agar overlay (WEBB et al., 1964). Recently FIALA and KENNY (1966) have used monolayers of rhinovirus-sensitive HeLa cells (Fig. 4) in a medium with additional magnesium and DEAE. Overlays of ion agar, agarose and methyl cellulose have also been used to solidify the overlay; normal agar may be inhibitory but the inhibition may be eliminated by addition of DEAE. However, the techniques are not easy and this has hampered the making of thorough going studies of the growth cycle of these viruses.

Haemagglutination of trypsin treated human red cells by rhinovirus 1A was reported by PELON et al. (1957). However, haemagglutination has not been observed by others (e.g. HAMPARIAN et al., 1961; MUFSON et al., 1965; TYRRELL et al., 1960) and in one case a rhinovirus preparation which agglutinated un-treated red cells was shown at Salisbury to be contaminated with SV5 (PHILIP-SON, unpublished).

It is possible however that rhinoviruses as a group react under certain specific circumstances with trypsin treated red cells. It was shown by DRESCHER and SCHRADER (1964) that red cells suspended in glucose buffered at about pH 5 agglutinate or aggregate. This haemaggregation is inhibited by sodium chloride and also by polioviruses, heated influenza viruses, adenoviruses, tuberculin and purified nucleic acids, both DNA and RNA. Some preparations of rhinoviruses do so also; the inhibition is not prevented by treatment with nucleases and al-though it is not sedimented at 80,000 g for 60 min. it is prevented by anti-serum and the phenomenon appears to be strain specific (TYRRELL et al., 1967).

The M rhinoviruses were first shown to grow in rhesus kidney cells, but will also multiply in similar cells from vervet, patas and cynomologus monkeys, both in primary or secondary cultures, and in some continuously cultivable monkey cell strains (Fig. 3). Some preliminary growth curve experiments were done with an M rhinovirus in rhesus monkey kidney cells (TYRRELL, 1963a). These showed that in cultures in optimal conditions the latent period was about 12 hours and that large amounts of virus were produced and quite rapidly shed into the medium. If the temperature was reduced from 33°C to 30°C the latent period was slightly prolonged but the virus was produced in large amounts and largely retained within the cells. These and other changes produced by varying the conditions are shown in Fig. 5. The reasons underlying the relatively specific effects of the different changes are not known but it may be significant that in studying the growth of several viruses in tissue culture cells it has been found that lowering the temperature to about 30°C hinders release (e.g. herpes and poliovirus) while increase of temperature above the optimum prevents virus synthesis (e.g. poliovirus and arboviruses).

The cytopathic effect of these viruses has been studied to some extent al-though much still remains to be done. Cells in unstained preparations can be

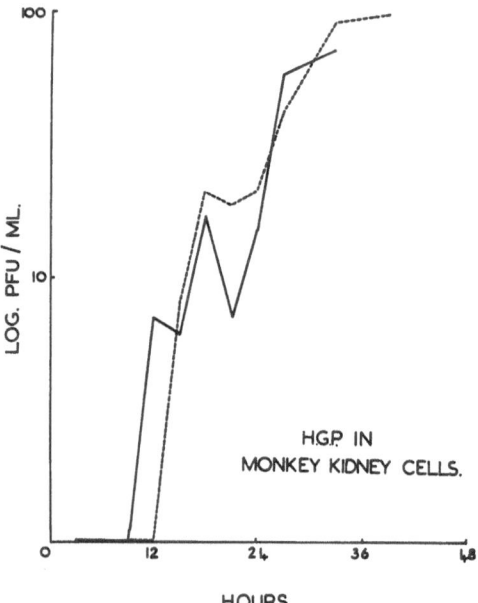

Fig. 5a – d. The multiplication of rhinovirus type 2 in roller tube cultures of monkey kidney cells under a) favourable and b), c) and d) unfavourable conditions. (TYRRELL, 1963 a. By permission of Hoeber.)

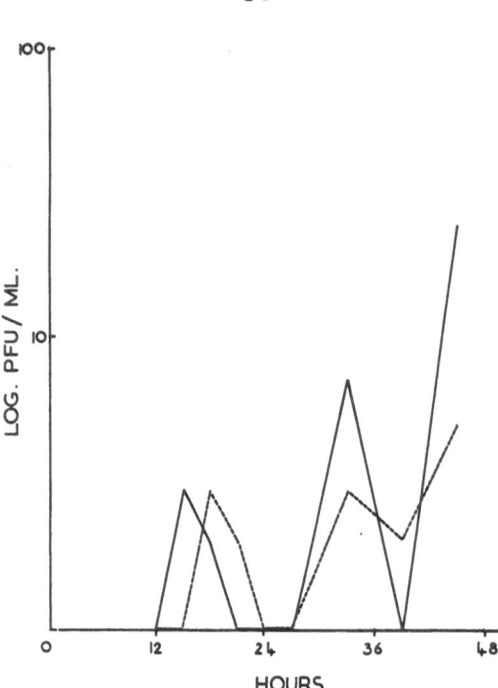

Fig. 5b

seen to draw away from one another. They tend to become irregular in outline and rounded up and refractile, the cytoplasm often appears granular and strands which end in irregular fragments of cytoplasm may be seen to extend from the surface of the cell (Fig. 6). Eventually the cells become completely rounded, some shrink or disintegrate into a mass of granules and others become detached and float away in the medium. The same major changes can be seen in cells which have been stained. These show also that at an early stage an eosinophilic area develops in the cytoplasms which seems to push the nucleus aside. Fragments of cytoplasm seem to be detached from the surface of the cell which then becomes irregular in outline, the pattern of chromatin in the nucleus becomes coarsened and eventually pyknotic (TYRRELL and BYNOE, 1961; TYRRELL, 1962, 1963a). Acridine orange staining shows an increased red staining of the cytoplasm of infected cells (TYRRELL, unpublished; DEIBEL and DUCHARME, 1965). Because the initial pathological changes take place in the cytoplasm and because of the analogy with enteroviruses it is tempting to believe that rhinoviruses replicate in the cytoplasm but this is not proven by direct experiments. Particles seen in electron micrographs of the cytoplasm of infected cells could not be positively identified as being virus particles (ARMSTRONG and TYRRELL, 1963, HAMPARIAN et al., 1961).

There is no information on many other aspects of rhinovirus

multiplication such as, for instance, whether the replication is inhibited by Actinomycin D and therefore whether it is dependent on DNA induced RNA. We should also like to know more about the synthesis of viral RNA and protein and possible precursors such as polymerase and other antigens. Such work depends on the development of a system in which definitive studies of one-step growth curves can be performed. It is possible that further modifications of the techniques developed by FIALA and KENNY (1966) may achieve this. Rhinovirus type 2 will multiply in HeLa cells in spinner cultures (STOTT and TYRRELL, 1967). Much further work is needed and the results might help to explain the means by which the adverse effects, such as change of temperature, inhibit the multiplication of these virus.

As mentioned earlier it has been shown that rhinoviruses multiply in organ cultures of human embryo tissues. A representative experiment with an M rhinovirus is shown in Fig. 7. This shows that virus is produced to the highest titre by nasal and tracheal epithelium. There is little or no growth in oesophageal and none in palatal epithelium. A few days after virus multiplication has reached its peak the ciliary activity which can be seen by reflected light is reduced, and sections of fixed tissue show that there is degeneration of the epithelial sheet.

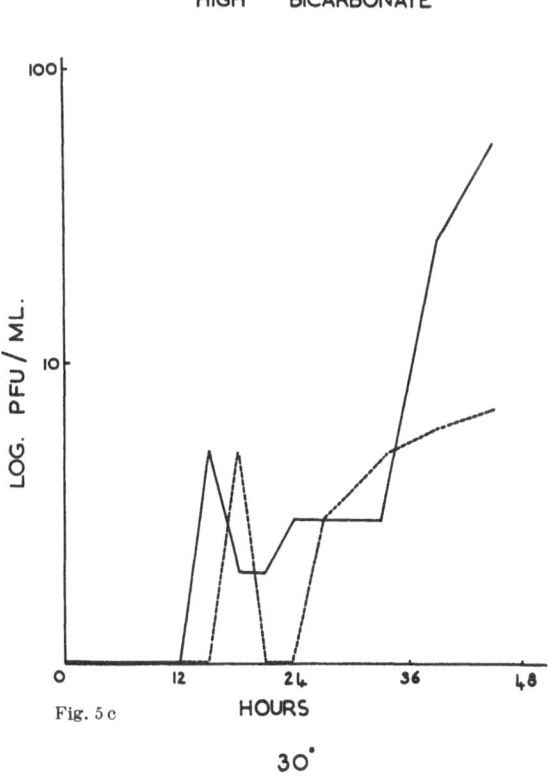

Fig. 5 c

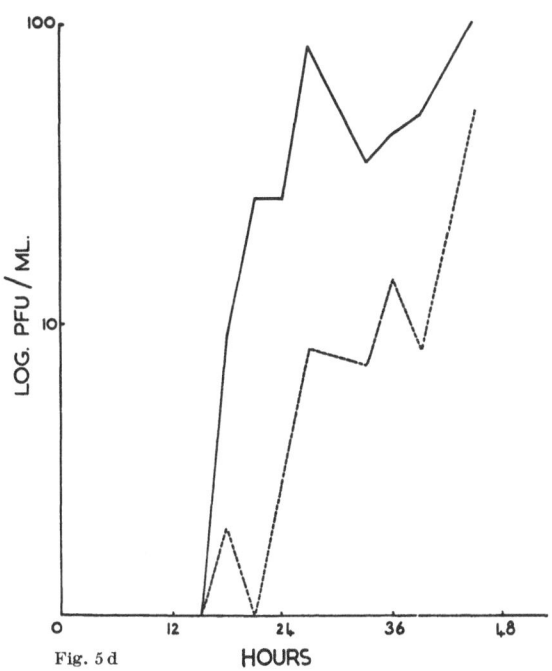

Fig. 5 d

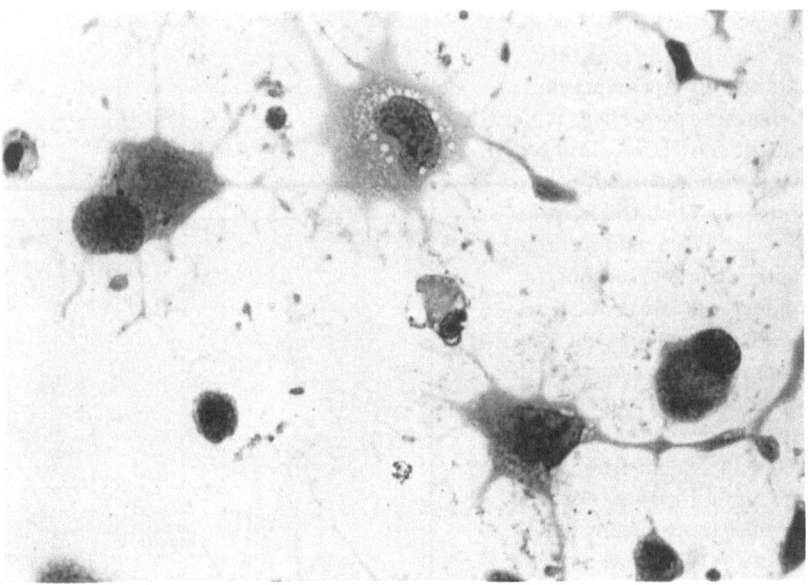

Fig. 6. Stained cultures of monkey kidney cells infected with an M rhinovirus (type 2) and showing moderately advanced degenerative changes and a typical eosinophilic cytoplasmic mass. Haematoxylin and eosin ×560. (TYRRELL and BYNOE, 1961. By permission Brit. med. J.)

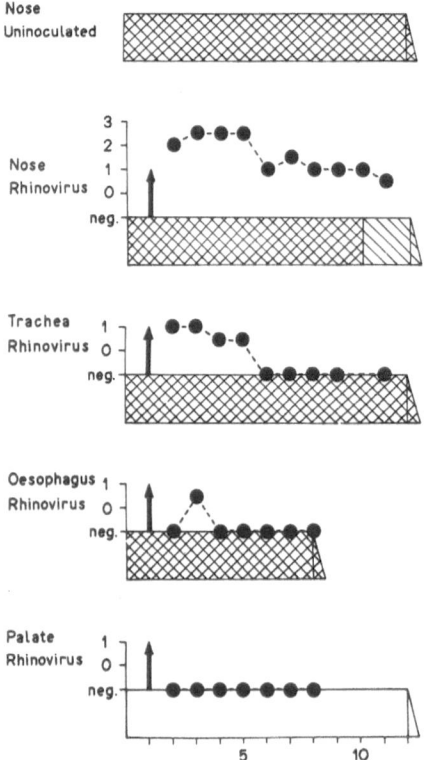

When washings from which no virus can be isolated are inoculated into organ cultures ciliary activity may cease and it may be presumed that a virus is present. In other cases after 1 or 2 passages inoculation of medium from the organ cultures induces the typical cytopathic effect of rhinoviruses in cultures of human fibroblasts although there may be no clear cut reduction in the ciliary activity in the organ culture (TYRRELL and BYNOE, 1965, 1966; TYRRELL and BLAMIRE, 1967). In other cases the cilia of organ cultures are damaged only after the virus has been adapted to the cultures by serial passage and even then it may produce only a low titre of virus. Rhinoviruses which produce a cytopathic effect when organ culture fluid is inoculated into tissue cultures may be adapted with

Fig. 7. The multiplication of rhinovirus type 2 in organ cultures of human embryo tissues. Virus grew to differing extents in different type of cells and not at all in palatal epithelium. (HOORN and TYRRELL, 1965. By permission Brit. J. exp. Path. and H. K. Lewis.)

difficulty to serial passage in fibroblast cells but it has nevertheless been possible to show that many of them have the essential properties of rhinoviruses (TYRRELL and BYNOE, 1966). Other rhinoviruses may be propagated only by serial passage in organ cultures and recognised by their destructive effect on cilia; they fail to multiply at all in other cells and may be called O strains. One such virus, known as HS, has been studied in some detail by the tests mentioned earlier, all measurements of infectivity being made by inoculating serial dilutions of virus into groups or organ cultures of human nasal or tracheal epithelium (HOORN and TYRRELL, 1966). The tests showed that it was ether-stable, acid-labile, and multiplied best at 33°C, with low concentrations of bicarbonate in the medium. Multiplication was not inhibited by BUDR, and the infectious particle passed a 50 mµ filter. It is therefore a typical rhinovirus and Fig. 8 portrays a growth curve in human nasal epithelium in which the latent period is evidently

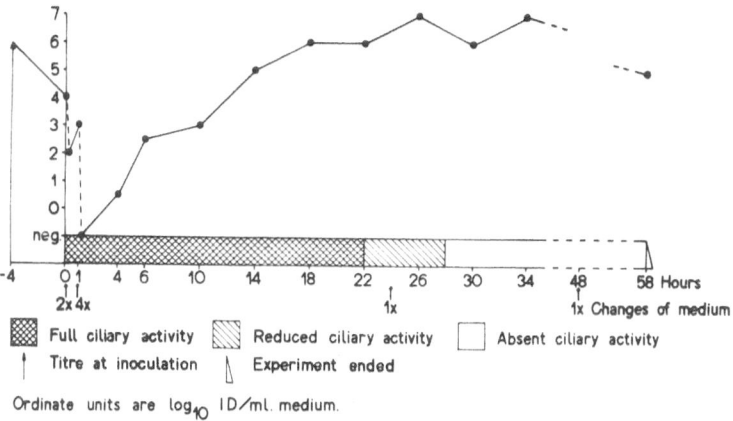

Fig. 8. The multiplication of the HS rhinovirus in organ culture of human embryo nasal epithelium. Adsorption was performed at 4° C for 4 hours. Incubation at 33° C started at 0 hours. Samples up to 6 hours after incubation were collected from the same dish but further samples were collected from a different dish each time. Virus was titrated in embryo trachea cultures using 2 dishes for each 10-fold dilution. (HOORN and TYRRELL, 1966. By permission Arch. ges. Virusforsch.)

under 12 hours. Fig. 9 shows that at about 18 hours after infection cells are beginning to be shed so that by 22 hours cliliary activity is impaired and gross changes are visible on section. Thereafter it is interesting to note that little more virus is produced and that deeper layers of cells are not attacked. Evidently this virus is extremely fastidious and multiplies only in fully differentiated epithelial cells and not in their precursors. There are again great gaps in our knowledge of how and when the virus multiplies and produces this cytopathic effect, but the effects themselves are probably very similar to those occurring in the nasal epithelium of patients with at least some sorts of common colds. In fact shed cells have been recovered from the culture medium and these are very similar to those which have been found in the nasal secretions in the early stages of a cold (Fig. 10). The organ culture technique developed by HOORN is thus not only a means whereby previously recalcitrant rhinoviruses can be persuaded to replicate within

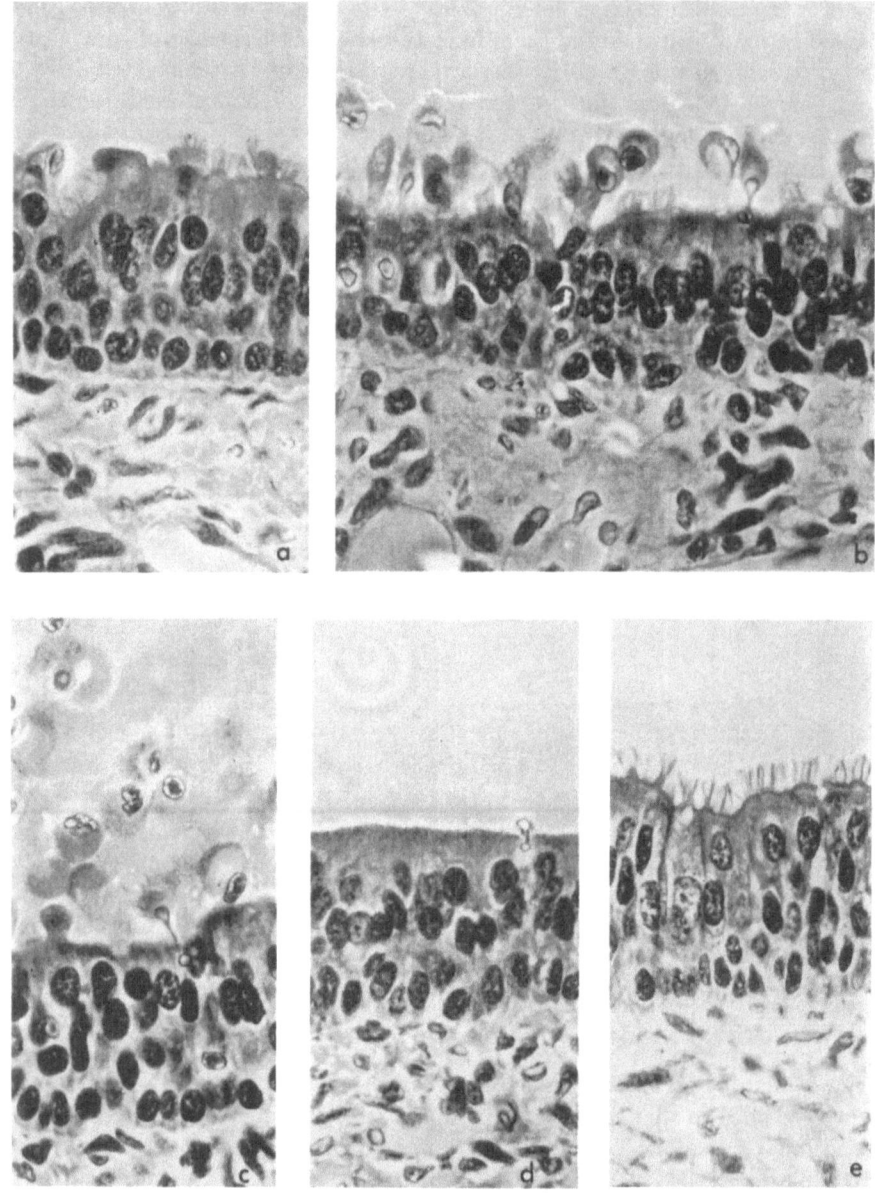

Fig. 9. The progress of degenerative changes in cultures inoculated with HS rhinovirus.

a) 18 hours after inoculation (a.i.). Well preserved cells except for single ciliated cells apparently
leaving the surface.

b) 24 hours a.i. Numerous rounded and vacuolating ciliated cells with pycnotic nuclei leaving the
epithelial surface.

c) 30 hours a.i. Still more shed cells and amorphous debris are in the medium and the remaining
superficial cells are irregularly shaped and appear somewhat disorganized.

d) 58 hours a.i. Almost all the shed cells have disappeared and a more regular epithelial surface has
now formed, but not ciliated cells are seen.

e) Uninoculated culture fixed at the same time as d).

All cultures were fixed with Bouin's fluid and stained with haematoxylin and eosin. Photographed by
phase contrast illumination (×360). (HOORN and TYRRELL, 1966. By permission Arch. ges. Virusforsch.)

the laboratory but also an elegant means whereby their effect on differentiated epithelial cells may be observed and studied.

There has been little published work on the detailed study of the multiplication of animal rhinoviruses; they seem to grow best under the optimal conditions for human viruses except that they are usually highly specific towards cells

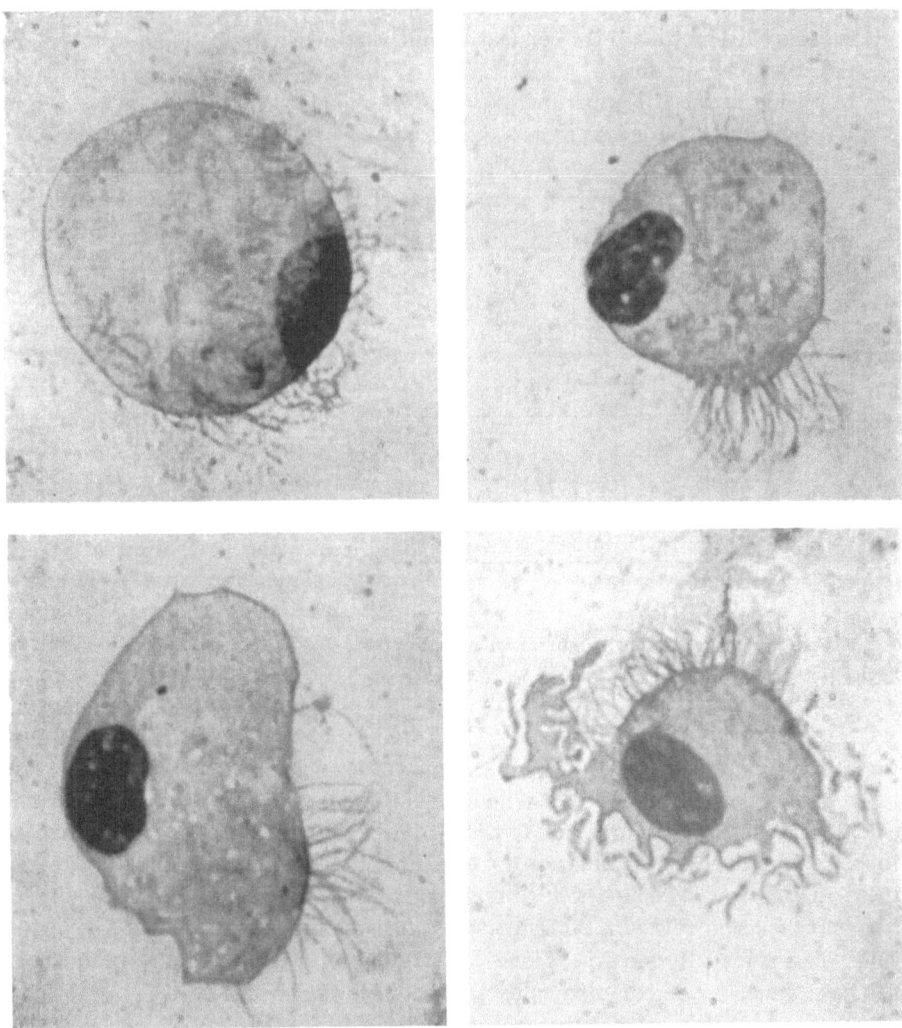

Fig. 10. A ciliated cell floating in the medium of a culture such as that shown in Fig. 9. Detached ciliated epithelial cells obtained from the medium of an infected culture of human embryo nasal epithelium one day after inoculation. Air dried, stained Giemsa (×800). (HOORN and TYRRELL, 1966. By permission Arch. ges. Virusforsch.)

derived from the animal in which they ordinarily grow (BÖGEL and BÖHM, 1962; BÜRKI, 1965). The exception here perhaps is the equine rhinovirus which seems to be able to infect man and human cells as well as the horse (PLUMMER, 1963; DITCHFIELD and MACPHERSON, 1965).

G. Pathogenesis

It is often necessary to study the pathogenesis of virus infections of man by analysing the course of events in a "model" infection in an experimental animal. Human rhinoviruses have not been found to cause disease in animals although those tested include all the common laboratory animals (PELON, 1961; HITCHCOCK and TYRRELL, 1960; JOHNSON and ROSEN, 1963). Nevertheless since it is ethical to infect volunteers with these viruses it has been possible to make controlled studies of the replication of these viruses in man and the relationship of this to symptoms.

In a short series of experiments virus infection was produced by swabbing a rhinovirus onto the nasal mucosa and conjunctiva but not on the throat. In the infection induced by the conjunctival route virus was found in the nose and not in the conjunctiva (BYNOE et al., 1961). Similar results have been obtained with Coxsackie virus A21 (BUCKLAND et al., 1965) and it is therefore probable that virus infection is initiated in nature by the attachment of a virus particle to cells of the nasal epithelium. This must be a rather remarkable phenomenon since most of the material deposited on the nasal mucosa is removed in a matter of minutes, apparently by being trapped in the mucous blanket and being moved into the pharynx by the action of cilia. Since rather small doses of virus seem to be able to initiate infection it is likely that the virus can nevertheless become attached to cells and initiate the process of infection. It has been found possible to infect volunteers with a fine aerosol of a rhinovirus (CATE et al., 1965). Administered in this way, virus particles impinge mainly in the lower respiratory tract and in correlation with this volunteers who became infected developed the symptoms of a tracheo-bronchitis. When a coarse spray is used, droplets impinge in the nasal mucosa and common colds are produced.

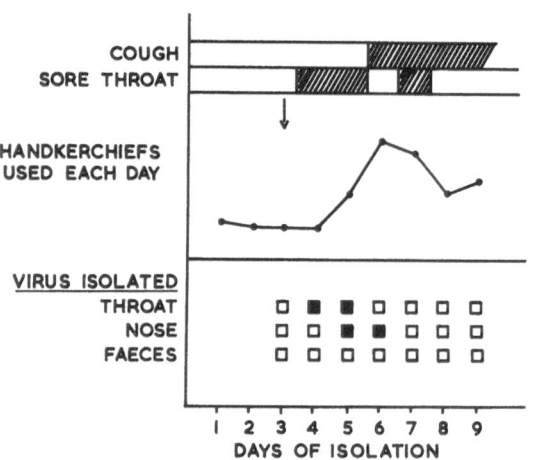

Fig. 11. The shedding of rhinovirus during an experimental cold. The virus was administered on the day shown by the arrow. Over 20 paper handkerchiefs were used daily at the height of the cold. (TYRRELL and BYNOE, 1961. By permission Brit. med. J.)

On the first day after intranasal inoculation it is often impossible to detect virus in the nose but on the second day virus is found in the nose and throat of susceptible subjects. At about this time the first symptoms are noted and the maximum concentration of virus in the nasal secretion is usually found when a profuse nasal discharge develops (TYRRELL, 1963a; DOUGLAS et al., 1966c). The concentration of virus then declines quite rapidly but may continue at a low level for some time. The peak concentration of virus is not closely related to the severity of the illness although DOUGLAS et al. (1966c) distinguish volunteers who excrete high titres of virus soon after infection ("early, high" group) and often have colds, from other volunteers, also without antibody who excrete little virus later on (the "late, low" group) and are less likely to get colds. Some representative results from our studies are shown in Fig. 11. In experiments at Salisbury small doses of virus in nasal washings have been given, while in the U.S.A. large doses of virus passed several times in tissue culture have been given (e.g. CATE et al., 1964). The general pattern of virus excretion is however, similar.

Table 3. *The Relation between Serum Antibody and the Response of Unvaccinated Volunteers to Experimental Infection with Rhinoviruses*

Reference	Virus inoculated	Titre of serum antibody before inoculation	Developing colds	Number of volunteers		Infected
				Excreting virus	Antibody rise	
1.	353 Type 13 Tissue culture fluid	4 or less	13/16	13/16	16/16	16/16
		8 — 16	6/12	10/12	12/12	12/12
		32 — 64	7/18	10/18	13/18	15/18
		128 — 256	3/12	7/12	8/12	9/12
		512 or more	0/9	1/9	1/9	2/9
2.	Thompson Type 6 washings	0.2^1 (less than)	11/11	11/11	6/10	11/11
		0.2 or more	0/13	3/13	5/11	6/13
3.	HGP Type 2 washings	0.22 (less than)	9/15	7/15	—	—
		0.22 or more	2/20	3/20	—	—
4.	HS organ culture fluid or washings	5 or less	4/9	9/9	6/9	9/9
		10 — 80	1/4	1/4	1/4	1/4

[1] K values of serum neutralizing activity.

1. MUFSON et al. (1963). 2. TAYLOR-ROBINSON and BYNOE (1964). 3. BYNOE et al. (1961). 4. HOORN et al. (1966).

Table 4. *Changes in the White Cell Counts of 17 Volunteers Infected with Rhinovirus Type 15 (Strain 1734)*

Category of volunteers	Cell type	White cell counts (1000/mm³)	
		before inoculation	2 days after inoculation
Ill	Neutrophil	3.83 (2.48 — 6.86)	5.88 (3.05 — 9.48)
	Lymphocyte	3.02 (1.61 — 4.23)	2.60 (1.44 — 4.03)
	Total:	7.46 (5.2 — 9.4)	9.35 (6.1 — 12.7)

There was no rise in inoculated volunteers who did not become infected or became infected and did not develop colds. Data from DOUGLAS et al. (1966) and by personal communication from Dr. R. G. DOUGLAS, Jr.

Five of 15 volunteers who developed rhinovirus-colds had an elevated ESR of 27 — 50 min/hrs. by the Westergren method (CATE et al., 1964).

In both types of experiment it has been shown that volunteers without antibody against the virus are usually infected and develop colds, those with high levels of antibody are neither infected nor develop colds, while those with intermediate levels may become infected but are less likely to develop colds than those without antibodies (BYNOE et al., 1961; MUFSON et al., 1963; CATE et al., 1964). The range of response to inoculation with these viruses is shown in Table 3. In one study blood counts and urine tests were performed and the results of certain of these are summarized in Table 4.

The mechanism by which virus is eliminated from the nose is not properly understood. Antibody secretion cannot be the explanation, for virus may disappear and the cold may recover in subjects who do not develop an antibody response, at least as judged by the microplaque reduction test titrations of the serum (TYRRELL, 1963a; TAYLOR-ROBINSON and BYNOE, 1964). It is possible that interferon is produced by infected cells and that this makes uninfected cells resistant and so limits the spread of infection. However, interferon has not so far been identified in the nasal secretions of subjects with colds although it has been found in the serum of patients with unspecified types of more severe respiratory disease (WHEELOCK and SIBLEY, 1964). Nevertheless it is quite likely that interferon does play a role for it has been shown that interferon treatment renders the cell of tissue cultures resistant to infection with a rhinovirus (SUTTON and TYRRELL, 1961), although this has not so far been shown with ciliated respiratory epithelium. There is a third possible explanation, namely that the susceptible ciliated epithelial cells are completely destroyed. Such destruction apparently renders the ferret insusceptible to infection with influenza virus by the nasal route, but we do not know that all the ciliated epithelium is destroyed by all viruses; it seems unlikely that this is so. Whatever the explanation, it is clear that there is a virus specific resistance to infection of volunteers with rhinovirus which is induced by a previous infection with such an organism which is not due to specific antibody and may persist for weeks following infection (CATE et al., 1964; FLEET et al., 1966). Here is another field for further research.

It has been mentioned that antibody against rhinoviruses may be found in human serum and that the presence of such antibody is correlated with resistance to infection. After experimental infection with small doses of an M rhinovirus there is usually an antibody response although the rises are less frequent after infection with an H rhinovirus (TAYLOR-ROBINSON et al., 1963b; TAYLOR-ROBINSON and BYNOE, 1964). It could be postulated that this antibody might find its way in the nasal secretion and there eliminate virus impinging on the mucosa. In support of this there is generally a correlation between the antibody titre of the serum and that of the nasal secretion (TYRRELL, 1963a); the latter titre has the same strain specificity as that of the serum and is apparently due to true γA or γG globulin antibody (ROSSEN et al., 1966a, b). Increased amounts of antibody are found in nasal secretion after vaccination (PRICE et al., 1959; TAYLOR-ROBINSON and DIMMOCK, 1964) and after rhinovirus infections (ROSSEN et al., 1966b). Immunological studies of volunteers infected with uncharacterised cold viruses show however that whether a cold develops or not a few days after inoculation some subjects begin to produce nasal secretion with an increased concentration of immunoglobulins, and since some of these are γG in type they may leak through an area of damaged epithelium too small to induce symptoms recognizable as a common cold (ANDERSON et al., 1962). This observation was not, however, confirmed by other observers. Probably both these mechanisms are important in the suppression of infection in an immune subject. Following infection, neutralizing antibodies against rhinoviruses persist for years (CONNELLY and HAMRE, 1964; HAMPARIAN et al., 1964a; SCIENTIFIC COMMITTEE ON COMMON COLD VACCINES, 1965; TAYLOR-ROBINSON, 1963a) although after a few years there is some decline in titre in many cases. In one study the titres against type 2 were unchanged (TAYLOR-ROBINSON, 1963a) over a period of years in one subject. Antibody responses are more frequent in volunteers and patients infected with M rhinoviruses than with H strains (TAYLOR-ROBINSON and BYNOE, 1964; TAYLOR-ROBINSON et al., 1963; HAMRE et al., 1964).

Finally, in spite of one unconfirmed result of PRICE and colleagues, it has been the general experience that rhinoviruses cannot be recovered from the faeces of infected patients or volunteers (TYRRELL and BYNOE, 1961; MUFSON et al., 1966b). This confirms the preference of these agents for the upper respiratory tract and, as mentioned earlier, presumably means that virus particles which are swallowed are either inactivated by the gastric acid, or fail to multiply in the gastrointestinal tract because the temperature is too high or because the cells are unsuitable. Since live virus vaccination with enteric-coated capsules does not seem to work well it is probable that the intestinal tract is a relatively unfavourable site for virus multiplication. It is interesting that the Coxsackie virus A21 which behaves in many ways like a rhinovirus can in fact be recovered from the faeces of both patients and volunteers although it seems to grow best in the respiratory tract (JOHNSON et al., 1962).

H. Variation

It is common experience that after many serial passages in tissue cultures recently isolated viruses begin to grow freely in cells in which they would not grow before. This, by analogy with other viruses, probably represents the selection of mutant rhinoviruses with a wider host range than the parent. However, no studies analysing the genetic basis of the phenomenon have yet been reported.

In addition, apparent mutants have been obtained by exposing infected cultures to guanidine. Wild rhinoviruses all seem to be unable to multiply in cultures maintained in a medium containing a low concentration of guanidine but if they are passed in increasing concentrations of the substance they first become resistant and may later become dependent. These results were obtained with one M and one H strain and the H strain appeared to be the more unstable. In similar experiments using a steroid antibiotic, fusidic acid, there was no change in sensitivity after 10 serial passages. The same viruses were passed in increasing concentrations of homologous antiserum and there was evidence of a slight change in antigenic structure which however developed much less readily than did changes in sensitivity to guanidine (DOGGETT et al., 1967).

Some strains of virus have been adapted to grow freely at 37°C (DRAPER and DIMMOCK, unpublished) and these may represent "hot" mutants. One of them is much less sensitive to urea than the parent strain (DIMMOCK, unpublished) and this may reflect the same phenomenon as the greatly increased urea sensitivity of "cold" mutants of polioviruses.

J. Immunitiy

The subject of immunity to rhinovirus infections has been touched on in earlier sections. In early phases of work on colds one could postulate that the illness was common either because it was due to infection with a few organisms to which there was no immunity or because it was due to infection with each of a large

number of organisms each of which produced immunity; since there were so many
different serotypes complete immunity was rarely achieved in a lifetime. It
now seems that the second alternative is more true than the first.

It was shown in experiments with uncharacterised nasal washings that volunteers who had been infected with one virus were apparently susceptible to infection with and to colds produced by any one of four others but resistant to re-
infection with the same washing (JACKSON et al., 1959, 1962). It was found
that pooled human globulin neutralized the cold-producing-capacity of nasal
washings and it was therefore fair to assume that there was a specific immunity
to colds mediated by antibody (JACKSON et al., 1958; TYRRELL et al., 1960).

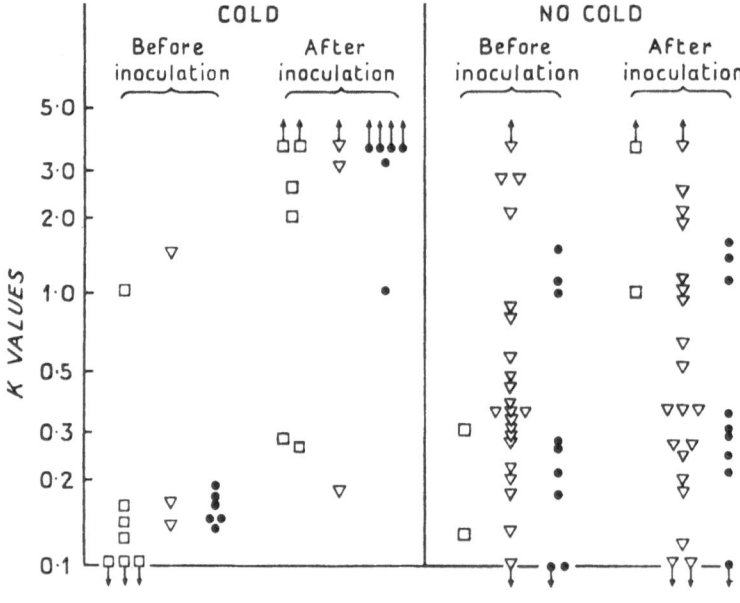

Fig. 12. The relationship between serum antibody levels and resistance to colds induced by rhinovirus
type 2. Antibody levels in the serum of volunteers before and about two weeks after inoculation.
(BYNOE et al., 1961. By permission Lancet.)

It was later shown that volunteers who carried serum antibody against
rhinoviruses were not infected when virus of the same strain was inoculated
intranasally (BYNOE et al., 1961). There was a clear correlation between an antibody activity (K) of 0.2 or greater and resistance to infection (Fig. 12), and
similar results were obtained in later experiments with H rhinoviruses, although
in both studies the presence of antibody did not regularly prevent infection
(TYRRELL et al. 1962; TAYLOR-ROBINSON and BYNOE, 1964); in some volunteers who where carrying antibody, rising titres were detected although there
were no symptoms and no virus was detected in the nasal secretion.

Using large doses of virus which had been passed in tissue cultures and administered as a spray it was found that volunteers with significant levels of antibody
developed colds and most were infected, but there was a clear effect of antibody

on the response to infection (Table 3). Volunteers with higher titres of antibody developed colds less frequently and shed less virus than those with a low titre of circulating antibody (MUFSON et al., 1963). The high titres of antibody required to prevent completely infection by such a heavy dose were in these experiments induced by intramuscular vaccination with a formalin inactivated vaccine. The strain used was a type 1A virus. Similar experiments were done in Britain with type 2 virus. It was shown that intramuscular injection of live or inactive virus gave an antibody response, whereas oral administration of live virus did not (DOGGETT et al., 1963). Intranasal administration only produced an antibody response if it induced infection and a common cold. As shown in Table 5 the vaccine used was prepared in monkey kidney tissue cultures. Two doses were given one week apart before exposure to live virus. Probably because of the small challenge dose used this vaccine completely prevented colds developing after intranasal administration of type 2 virus, but over one-third of vaccinated volunteers developed colds when inoculated in the same way with a strain of type 1A virus.

Table 5

Results of Inoculation of Volunteers Previously Given Inactivated Rhinovirus Vaccine

Reference	Vaccine given	Homologous antibody response to vaccine	Challenge virus	No. of colds	No. of virus isolations	No. of antibody rises
1.	Type 2	23/27	Type 2 washings	1/28	5/18	11/27
	None	—	Type 2 washings	11/23	14/23	15/22
	Type 2	12/13	Type 1A washings	6/13	11/13	9/13
2.	Type 1A	22/28	Type 1A tissue culture fluid	11/22[1]	19/22	17/22
	None	—	Type 1A tissue culture fluid	2/6	5/6	6/6

[1] Only 1/9 volunteers with antibody titres of 32 or greater developed colds.
1. Scientific Committee on Common Cold Vaccines. 2. MUFSON et al. (1963).

Solid protection of this sort against challenge with a rhinovirus is of scientific interest but little practical value. It would obviously be desirable to be able to vaccinate with a polyvalent vaccine but in order to be able to do this it would be necessary to be able to obtain a good antibody response following the injection of very small doses of rhinovirus. It is not feasible at the moment to manufacture and combine highly concentrated preparations of many different rhinoviruses. Attempts were therefore made to produce an adjuvant by emulsifying vaccine with an oil adjuvant and by adsorbing virus onto aluminium phosphate; neither of these methods was successful (SCIENTIFIC COMMITTEE, 1965). It is clear that many more studies are needed before rhinovirus vaccines are likely to be successful. For instance, although small amounts of vaccine have been made with H viruses grown in human diploid fibroblast cell strains it would be necessary to improve the techniques so that higher titre material could be obtained and be purified, perhaps by zonal ultracentrifugation, and then combined with other strains for incorporation in a polyvalent vaccine.

There are other approaches to the problem of vaccination. The cross reactions between strains mentioned earlier might make it possible to boost the antibody response of subjects, particularly those who had already been exposed to some of the viruses prevalent in the community. On the other hand, it seems quite likely that different viruses may be found in different areas; if some epidemiological pattern could be discerned in this it might be possible to produce a vaccine for the inhabitants of a particular area, which included only those serotypes to which they were likely to be exposed. Alternatively, one might survey the antibody of a particular population and vaccinate them against those viruses to which they had no antibody; this might be reasonable if it could be shown that when a population was deficient in antibody against a virus, perhaps recovered in another part of the world, an epidemic with that virus was likely to occur. As will be explained below our knowledge of epidemiology is far too fragmentary to us to be able to speak from knowledge about any of these alternatives at the moment.

K. Relationship between Rhinoviruses and Disease

Rhinoviruses of man, as has been said before, were first recognised because they produced colds when administered to volunteers. It was of course realised that the colds might be due to some other virus with which the volunteers had been infected owing to failure of isolation or some other fault in the technique; but in many studies up to half of a group of volunteers were inoculated with "dummy" material of some sort and did not develop colds. It was also possible that the illness might be due to some virus other than that which could be recognised by inoculation of tissue cultures; this was unlikely because it was usually found by laboratory tests that there was a higher frequency of infection in those who developed colds than in those who did not, and that this difference was statistically significant. In addition those who possessed specific antiviral antibody were usually protected from illness. Finally it was possible that in experiments in which tissue culture fluid was used to induce colds the illness was due to some unrecognised virus derived from the inoculum or from the tissue cultures used. However, colds were not produced by medium from culture vessels which did not contain susceptible cells, or by medium from uninoculated cultures. Furthermore colds were produced by virus which had been passed serially a number of times in volunteers or cultures. It could be concluded that the virus really produced the colds observed. Virtually every virus tested so far has produced colds when administered to volunteers, and many of the results are summarised in Table 6.

It can be concluded that rhinoviruses administered intranasally to volunteers cause colds, but it is also necessary to study the relationship between rhinoviruses and naturally occurring disease. If rhinoviruses are responsible for naturally occurring colds it should be possible to detect infection more frequently in patients with colds than in those without. It should also be possible to look for virus infection in patients suffering from a wide range of respiratory infections, and possibly of other diseases as well, and so to determine whether in subjects other than healthy volunteers the virus could produce different or more serious types of disease. A good deal of work has been done already on these lines. Various populations have been studied and the results are summarised in Table 7. This shows that there is evidence that rhinoviruses are associated

Table 6. *Results of Inoculating Rhinoviruses into Normal Volunteers*

Reference	Serotype	Source of inoculum	No. of volunteers with			No. of colds in		
			Colds	Virus isolation	Antibody rise	Infected volunteers	Uninfected volunteers	Uninoculated volunteers
1	1A	Tissue culture fluid (14 passages) and nasal washings	6/58	11/58	0/24	4/11	2/47	
2	1A	Nasal washings	8/23	19/23	16/22	8/20	0/3	
3	1A	Tissue culture fluid (1 passage)	2/6	5/6	6/6	2/6	0/0	
4	1B	Nasal washings	4/11			3/3	1/7	1/26
5	2	Nasal washings	10/30	9/30		8/11	2/19²	
6	3	Nasal washings	11/24	14/24	14/21			1/27
	4	Nasal washings	4/8	5/8	3/8			
	6	Nasal washings	4/14	6/8	2/14			
3	13	Second passage of virus in tissue cultures of WI-26 cells	29/67	41/67	50/67			0/10
7	15	Second passage in WI-26 cells	23/38	28/38	31/37	23/33	0/5	
8	DC	Nasal washings	4/10	6/10	4/10	4/7	0/3	0/17
9	HS	Culture fluid (6) and nasal washings (7)	5/13	10/13	8/13	5/10	0/3	
10	Various H strains	Nasal washings or tissue culture fluids	19/50	22/50	N.D.	19/22¹	9/28	1/49

¹ Virus isolation only.   ² Serum antibody rise only.

References: 1. TYRRELL and BYNOE (1958). 2. Scientific Committee on Common Cold Vaccines (1965). 3. MURSON et al. (1963). 4. TYRRELL (1963a). 5. BYNOE et al. (1961). 6. TAYLOR-ROBINSON (1964). 7. CATE et al. (1964). 8. TYRRELL et al. (1962). 9. HOORN et al. (1966). 10. TYRRELL and BYNOE (1961).

Table 7. *Results of Attempts to Isolate Rhinoviruses from Patients and Healthy Persons*

Reference	Subjects tested				Patients				Control subjects		
	Age group	Clinical condition	Time	Place	Number tested	Virus isolated Number	Virus isolated %	Number tested	Virus isolated Number	Virus isolated %	
1.	Young adults	Common colds	1960	Sheffield, England	25	8	32	8	0	(0)	
2.	Students	URI	1961—64	Chicago, Ill.	894	227	25.4	2130	31	1.5	
3.	Children	URI	1961—64	Philadelphia, Pa.	508	28	5.5	155	4	2.6	
	Industrial workers	URI	1961—64	Philadelphia, Pa.	567	59	10.4	178	2	1.1	
4.	Children	Febrile or afebrile	1961—62	Camp Lejeune, N.C.	597	29	4.9	641	14	2.2	
	Recruits	respiratory disease			643	115	18	732	40	5.5	
	Seasoned Soldiers				514	35	6.8	453	7	1.5	
5.	Industrial workers	Respiratory illness	1963—66	Charlottesville, Va.	1025	239	23.3	255	4	1.6	
6.	Children	URI	1961—64	England	1321	49	3.7	493	10	2.0	
	Adults	URI	1961—64	England	467	29	6.2				
7.	Recruits	Acute febrile respiratory disease	1959—60	Parris Island	169	43	25	117	10	8.5	
8.	Children and adults	URI at home	1963—65	New York, N.Y.	1724	—	4.5	4205	—	0.9	
9.	Children	Respiratory disease in hospital	?—1964	Washington, D.C.	1826[2]	—	3.4	—	—	1.7	
10.	Children	Respiratory disease in hospital		Scotland	113	10	8.8	113[1]	10	8.8	

| 11. | Adults | Relapses of chronic bronchitis | 1964–65 | Sheffield, England | 56 | 8 | 14.3 | 237 | 5 | 2.3 |
| 12. | Adults | Pneumonia in hospital | | Washington, D.C. | 361 | 6 | 1.7 | 908 | 6 | 0.7 |

¹ Cases of diarrhoea. ² Cases and controls.
1. HOBSON and SCHILD (1960). 2. HAMRE et al. (1967). 3. HAMPARIAN et al. (1964). 4. BLOOM et al. (1963). 5. GWALTNEY et al. (1966). 6. Working Party (1965). 7. JOHNSON et al. (1962). 8. ELVEBACK et al. (1966). 9. CHANOCK and PARROTT (1965). 10. STOTT et al. (1967). 11. STENHOUSE (1967). 12. MUFSON (unpublished).
There have also been a number of thorough studies, e.g. HIGGINS et al. (1963, 1964), MUFSON et al. (1966), PEREIRA et al. (1963), EADIE et al. (1966) and KENDALL et al. (1962) in which control subjects were not tested but comparable isolation rates were obtained from patients.

with colds occurring in students and similar populations of young adults. It is also apparent that they may be recovered from a small proportion of normal persons — some of whom may be undergoing a symptom-less infection while others may be still shedding virus acquired at the time of a previous cold. It has been shown in the course of these studies that successive colds may in fact be due to infection with a series of immunologically, and sometimes biologically different viruses, to each of which in turn antibody and therefore presumably immunity is acquired. An example is shown in Table 8. There is little evidence that rhinoviruses are important in the causation of severe lower respiratory disease of children and none that they are important in diseases such as pneumonia of adults. The latter finding, however, may merely mean that the virus which initiated a mild respiratory catarrh can no longer be found after that mild illness has been made severe by infection with a pathogenic bacterium.

Rhinoviruses have been recovered from patients with acute exacerbations of a state of chronic bronchitis (EADIE et al., 1966; STENHOUSE, 1967). These patients sometimes had no upper respiratory tract disease but this may have been due to their breathing through the mouth because of their dyspnoea, so that the virus infection was initiated in the bronchial tree. The bronchi are certainly infected for titration of sputum may show concentrations of rhinoviruses in excess of 10^3 TCD$_{50}$/ml (STOTT, personal communication). Fuller studies of this problem are needed. In addition rhinoviruses have been recovered from two child patients with an unusual form of encephalitis accompanied by retinal haemorrhages (HOLZEL et al., 1965b). For some reason viruses were not isolated by these workers from any patients with respiratory disease, but the cases reported suggest the possibility that rhinoviruses may occasionally behave like enteroviruses and produce viraemia and infect organs distant from the respiratory tract; the possibility has not been seriously investigated, but it is likely to occur much more rarely than with enteroviruses.

Finally, the type of illness diagnosed in volunteers and patients shown to be infected with rhinoviruses is summarised in Table 9 and Fig. 13a and b.

Although the pathology of even the commonest type of human disease due to rhinoviruses has not been studied, the organ culture experiments described earlier can be regarded as a guide to what happens in nasal epithelium. HILDING (1930) studied biopsy and other material from uncharacterised colds, many of which may have been due to rhinoviruses. He observed oedema of the submucous tissue at the time of onset, followed by shedding of the surface epithelium which was completed by the third day of the disease and progressed to reach the deepest cell layers by the fifth day. During the first three days epithelial cells were found in the nasal secretion. The epithelium was completely regenerated by the fourteenth day from onset.

Table 8. *The Result of Testing for Rhinoviruses in Successive Colds in a Chicago Student* (HAMRE *and* PROCKNOW, *1963*)

Date of specimen collection	Result of test for rhinovirus
1960 October 4	0
1961 January 3	0
February 15	+
April 17	+
November 6	+
November 18	+
1962 February 5	0
March 27	0
April 17	+

All the rhinoviruses were antigenically distinct and were H-type strains.

Somewhat similar observations were reported by BRYAN and BRYAN (1953) who examined nasal smears stained by the Papanicolau technique. They also observed inclusion-like bodies in the cytoplasm.

It is still widely believed that the catarrhal stages of a cold are due to bacterial infection following the invasion of a virus. This is still unproven and in fact it could well be a reflection of the time taken by the epithelium to regenerate and recover its full function. On the other hand mycoplasmas would not be detected by standard bacteriological techniques and these organisms might multiply in or on the damaged epithelium and retard the healing process. This possibility could be investigated by the application of present techniques. Nevertheless bac-

Table 9. *Range of Illness Caused by Two Types of Rhinoviruses in Volunteers (*BYNOE *and* RODEN — *unpublished)*

Type and strain	Number inoculated	No. of colds	No. of colds of indicated severity			Mean	
			Mild	Moderate	Severe	Incubation period (days)	Duration (days)
HGP and PK	213	78 *37%*	63 *81%*	12 *15%*	3 *4%*	2.1	9.0
DC	251	77 *31%*	36 *47%*	28 *36%*	13 *17%*	2.1	9.9

The viruses were administered as nasal containing diluted nasal washings.

terial invasion may often be responsible for cases of profuse purulent nasal catarrh following colds, especially in children, and also for typical cases of acute sinusitis and otitis media.

L. Epidemiology

The results which were described earlier in support of the view that rhinoviruses cause respiratory disease were obtained in long-continued studies which also yielded information on the overall epidemiology of these viruses.

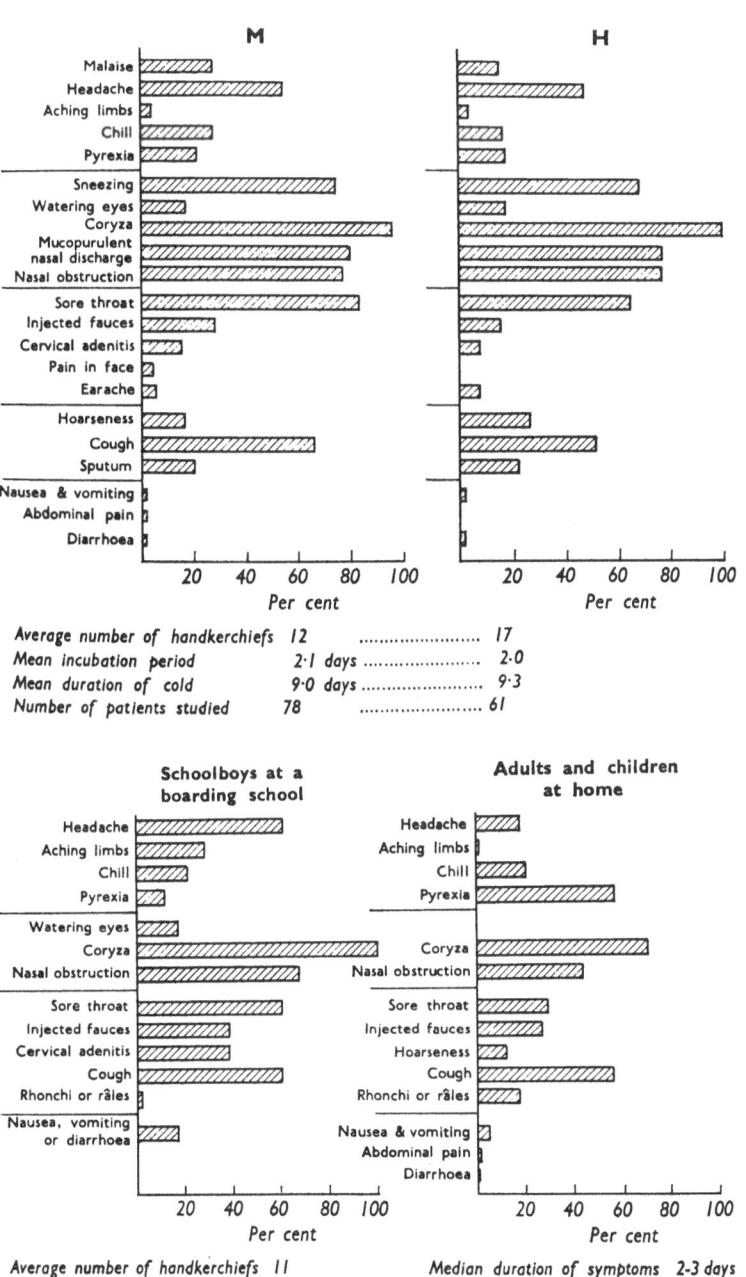

Fig. 13a, b. Clinical observations on a) volunteers and b) patients suffering from colds and other illnesses produced by rhinoviruses. (From TYRRELL, 1965. By permission Arnolds.) In the case of the volunteers the M rhinovirus was type 2 and the H rhinovirus was the strain DC.

Rhinoviruses can be recovered from patients of all ages from infancy to advanced old age. It is not easy to find figures to indicate whether infection is commoner at one age than another; some are given incidentally in Table 7. Most workers report finding rhinoviruses more commonly in the winter when respiratory infections are common. However, on several occasions it has been observed that, using tissue cultures, viruses could not be observed in the period around January and February (KENDALL et al., 1962; HAMRE et al., 1967),

Table 10. *Frequency of Colds in Volunteers Inoculated with Rhinoviruses, 1957—1966*

	January — March	April — June	July — September	October — December	Total
Number inoculated	159	237	148	65	609
Number developing colds	47	72	52	20	191
Percentage	29.5	30.3	35.1	30.8	31.4

This Table was kindly prepared by Dr. M. L. BYNOE. Most of the volunteers were inoculated with rhinovirus type 2 (strains HGP or PK) and smaller numbers with types 1, 9, 3 or 4 (strains JH, DC, FEB and 16/60). The apparent slight increase in frequency of colds in July — September is not statistically significant.

Table 11. *Frequency of Neutralizing Antibodies against Rhinoviruses in Sera Collected from Normal Adults in Various Parts of the World.* (TAYLOR-ROBINSON, *1965*; BROWN and TAYLOR-ROBINSON, *1966*)

	Origin of sera	Number tested	Number and percentage of positive sera against			
			Type 1A		Type 2	
Urban	Australia	19	18	95	15	79
	Jamaica	15	12	80	10	67
	India	15	13	87	12	80
	Malaya	15	13	87	11	73
	South Africa	15	11	73	9	60
	France	11	5	45	3	20
	England	22	16	73	13	59
	Chile	15	12	80	12	80
	Czechoslovakia	15	8	53	7	47
	U.S.A.	9	6	67	4	44
	Lebanon	15	11	73	10	67
Isolated communities	Hottentots (South Africa)	45	41	91	27	60
	Eskimos (North America)	54	46	85	31	58
	Micronesians (Pacific)	65	32	49	26	39

while others have seen epidemic waves of a rhinovirus at that time (MUFSON et al., 1966a). In one study it was found over a period of years that the fraction of all patients studied who yield rhinoviruses was higher in the summer than the winter (HOPE-SIMPSON, 1966). This suggests that H and M rhinoviruses have a seasonal incidence which is roughly the same as that of respiratory diseases in general but that respiratory virus diseases in mid-winter may also be due to viruses which cannot be cultivated by present tissue techniques. This conclusion, however, does not explain why the incidence fluctuates with the seasons. The best analysis so far was made by calculating partial correlation between the incidence of colds and many different measurable parameters of the weather. The results so far indicate that an increase in incidence of colds occurs 2—3 days after a drop in temperature, and that there is a weaker association with humidity (LIDWELL et al., 1965). This suggests that a drop in temperature influences in some way the transmission of the virus. There is no evidence at the moment that experimental exposure to a rhinovirus during the colder seasons of the year is more likely to induce a cold than exposure during the warmer months (Table 10). Subjects have been chilled and then exposed to uncharacterised viruses administered as nasal drops and generally they were no more likely to develop colds than unchilled volunteers (ANDREWES, 1951). The Chicago group did however produce an increased number of colds in women who were chilled in the middle third of the menstrual cycle (JACKSON et al., 1960b). It is possible that different results would be obtained if virus was administered as airborne droplets instead of in nasal drops but such experiments have not been reported.

The seasonal effect might occur because airborne droplets of virus survive better in the indoor environment in winter when the relative humidity has been reduced by artificially heating the air. HEMMES et al. (1960) have suggested that influenza is prevalent in winter because this virus survives well at low relative humidity while poliovirus is prevalent in summer because it survives better in air of high humidity. However other respiratory viruses do not behave like influenza. Virus on a glass slide loses its infectivity in much the same way as virus in airborne droplets and it has been shown in this system that rhinoviruses behave like other picornaviruses and are better preserved at high humidities (BUCKLAND and TYRRELL, 1962). This suggests that the humidity effect is not the determining factor in the seasonal fluctuation; it also suggests incidentally that infectious virus is not airborne for a long time during the normal transmission cycle, otherwise the effect of low humidity in winter would reduce transmission at that time of the year. It might however explain the observation in some studies that rhinoviruses are not common January and February in the Northern Hemisphere when external temperatures and therefore internal relative humidities are at their lowest. The seasonal effect which causes an increase of colds in the autumn might therefore occur in the patient transmitting infection either by increasing the amount of virus in the nasal secretions or by increasing the likelihood that it will be expelled as airborne droplets. Infection with Coxsackie virus A21 is in many ways like that seen in colds produced by a rhinovirus, and unpublished results (BUCKLAND, 1963) indicate that volunteers infected in cold weather shed more nasal secretion on coming indoors than do volunteers kept inside in a uniformly warm environment. It is in fact well known that the nose tends to run and require blowing more often in cold than in warm weather. It has been shown that blowing the nose is an effective way of producing an aerosol of particles likely to be trapped in the nose (BUCKLAND et al., 1965). The increased transmission of colds in cold weather might therefore be due to this winter dampness on the nose. However, by no means has the last word been said on the seasonal factors involved in respiratory tract infection; the syndrome of "shipping fever" of cattle shows how crowding and "stress", and changes in diet may be the determining factors in inducing an epidemic of respiratory disease and it is likely that each of them plays some part (ANDREWES, 1964). Some workers believe that common cold viruses are commonly present in a latent form in the population and that low temperatures activate these infections in some ways so that symptoms develop. Studies on rhinoviruses have so far given no experimental support for this view.

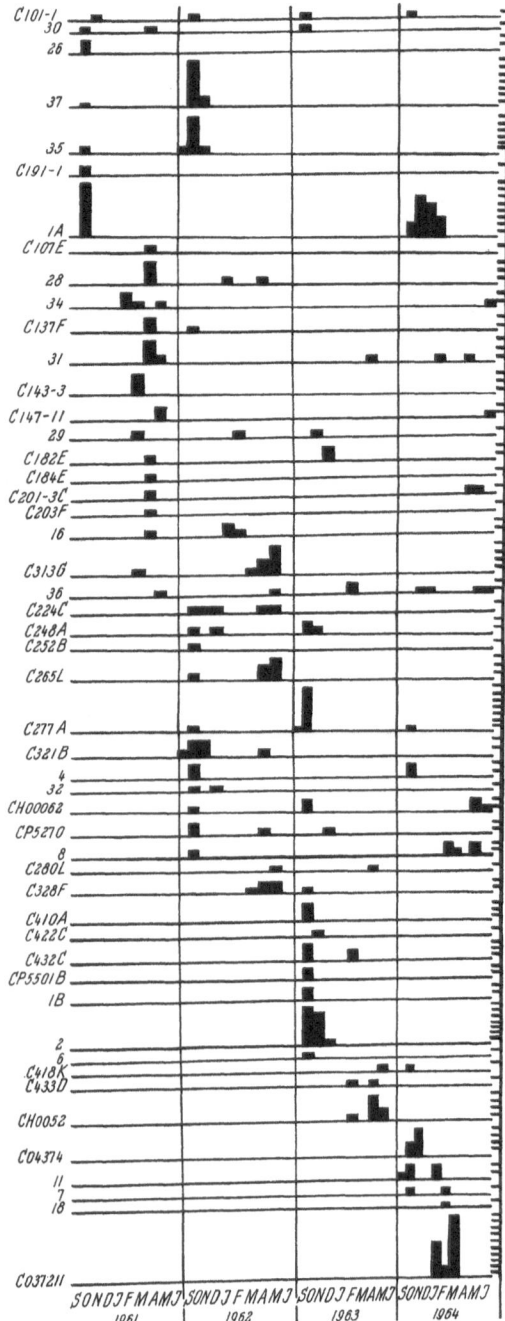

It is not possible to describe fully the distribution in space and time of the occurrence of the various serotypes of rhinoviruses. Nevertheless the general pattern of their behaviour is already becoming clear. Some representative data are given in Fig. 14. This makes it clear that there is a constant flux in the serotypes infecting any population — one type enters a community and persists for a matter of weeks or a few months and then entirely disappears. Also, in one study in a school it was shown that two serotypes of virus had disappeared without inducing antibody in more than a minority of the children (KENDALL et al. 1962). It appears therefore that even in a closed community there must be a rather high proportion of susceptibles if a rhinovirus is to continue to spread within it, and that the virus often fails to spread before all the susceptibles are exhausted. Furthermore it seems to be normal for several serotypes of rhinovirus to be circulating at about the same time. The winter respiratory "season" therefore appears to be formed by the summation of an overlapping series of epidemics induced by a succession of new viral serotypes, many of which are rhinoviruses.

Fig. 14. Incidence of Rhinovirus Serotypes, 1960—1964. The occurrence of virus of various serotypes in a population of young adults in Chicago. Redrawn from HAMRE et al. (1967).

Serological surveys are useful adjuncts to the study of the epidemiology of rhinoviruses. Several have been reported. It appears in general that maternal antibody is lost rather early and then antibody to one virus after another is acquired, often beginning in early childhood. Antibody is found in a substantial proportion of sera only in adolescence, although some studies indicate that there may be antibody against one serotype in the sera of many children while there are few with antibody against another (Fig. 15). This suggests that the interval since the last occasion on which a particular virus was prevalent in a community may vary over a range of time as great as that of human childhood. Therefore it is likely that certain serotypes may revisit an area at intervals of up to ten years or so. This is consistent with the data obtained so far on virus isolations which only extends over a period of four or five years — virus serotypes were rarely recovered twice on an epidemic scale in any one area. Serological surveys against a few M rhinoviruses have shown that antibodies against these are common in the sera of adults living in many parts of the world, and therefore the viruses are presumably circulating there too (Table 11). Even remote communities living in Alaska, the Pacific Islands, the Kalahari desert or the island of Tristan da Cunha carry antibodies against rhinoviruses (BROWN and TAYLOR-ROBINSON, 1966; TAYLOR-ROBINSON and TYRRELL, 1963) and these are compatible with the clinical observation that colds do occur in such communities, though they may be observed as distinct epidemic waves

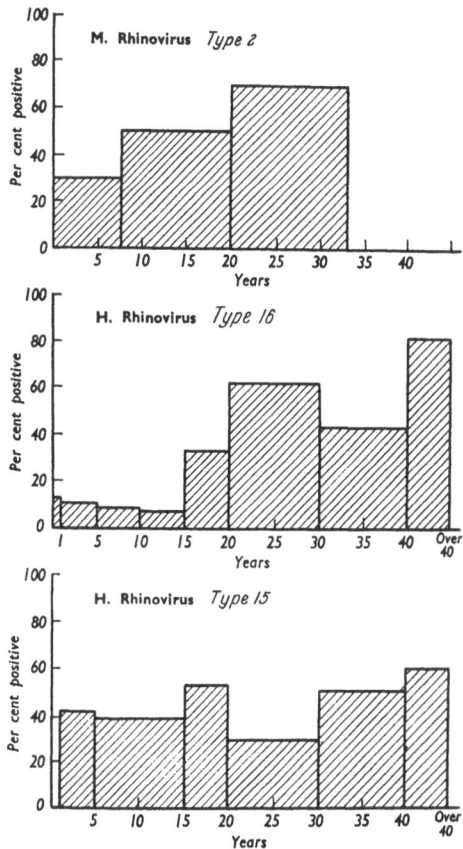

Fig. 15. The frequency of antibodies against rhinoviruses in subjects of different ages. Antibodies against type 2 were measured in the sera of subjects from Britain and against types 15 and 16 in subjects from the U.S.A. (TAYLOR-ROBINSON and TYRRELL, 1962b; JOHNSON and ROSEN, 1963. Figure from TYRRELL, 1965. By permission Arnolds.) See also SCHILD and HOBSON, 1962.

rather than as a prolonged season of increased prevalence. Rhinoviruses are therefore apparently ubiquitous infectious agents.

The above interpretations are probably sound but they are based on two assumptions, namely that antibody is induced by infecting with the virus used in the test or with one very much like it, and that new serotypes are not being selected by the passage of virus in the respiratory tract of humans having low titres of antibody. Analogy with enteroviruses suggests that if new serotypes are appearing they are not doing so rapidly, but experience with Foot-and-Mouth

Table 12. *Proposed Rhinovirus Numbering System (*KAPIKIAN *et al. 1961)*

Rhinovirus No.	Prototype strain[1]	References describing indicated strains
1A	ECHO 28	11, 12
1B	B632 [K779]	13, [14]
2	HGP	13
3	FEB	13
4	16/60	13
5	Norman	13
6	Thompson	13
7	68-CV 11+	15
8	MRH-CV 12	15
9	211-CV 13	15
10	204-CV 14	15
11	1-CV 15	15
12	181-CV 16	15
13	353 [5,007-CV 23]	16, 17, [18]
14	1,059	16, 17
15	1,734	16, 17
16	11,757	16, 17
17	33,342	16, 17
18	5,986-CV 17	18
19	6,072-CV 18	18
20	15-CV 19	18
21	47-CV 21	18
22	127-CV 22 [203 F]	18, [19, 6]
23	5,124-CV 24 (100,319)[1]	18, (9)
24	5,146-CV 25 [147H]	18, [19, 6]
25	5,426-CV 26 (K2,218)[1] (55,216)[1]	18, (20), (9)
26	5,660-CV 27 (127-1)[1]	18, (19, 6)
27	5,870-CV 28	18
28	6,101-CV 29 (113E)[1]	18, (19, 6)
29	5,582-CV 30 (179E)[1]	18, (21, 6)
30	106F	21, 6
31	140F	21, 6
32	363	22
33	1,200	22
34	137-3	19, 6
35	164A	19, 6
36	342H	19
37	151-1	19, 6
38	CH 79[2] [201-3C]	1, 7, [19, 6]
39	209 [00052]	23, [6]
40	1,794 [184E]	23, [19, 6]
41	56,110 [137F]	23, [19, 6]
42	56,822 [248A]	23, [6]
43	58,750 (E2 No. 133)[1] (WIS 258E)[1] [04374]	23, (20), 24), [6]
44	71,560	23
45	Baylor 1 (037211)[1] (E2 No. 46)[1]	25, (6), (20)
46	Baylor 2 [477-CV 50] [CH 202[2]]	25, [26], [1]
47	Baylor 3 [1,979M-CV 46] [CH 310[2]]	25, [26], [1]
48	1,505	9
49	8,213	9
50	A2 No. 58	20
51	F01-4,081 (19,143)[1] [605-CV 45] [313G]	27, (9), [26], [19,6]

(Continued Table 12)

Rhinovirus No.	Prototype strain[1]	References describing indicated strains
52	F01-3,772 (16,413)[1] [515-CV 34]	27, (9), [26]
53	F01-3,928 [252B]	27, [6]
54	F01-3,774 [2,253-CV 49]	27, [26]
55	WIS 315E [Baylor 4]	24, [28]

[1] Virus in parentheses represents a virus submitted to programme by a collaborating laboratory and found to be identical to the prototype strain; virus in brackets represents a virus not included in first phase or not submitted to programme and found to be identical with the prototype strain by a collaborating laboratory. Reference numbers are shown in a similar manner to above. Thus, reference number not in parentheses refers to prototype strain, number in parentheses refers to virus strain in parentheses. Number in brackets refers to virus strain in brackets.

[2] CH 79 was formerly designated as CHV/2/59; CH 202 formerly designated as CHV/7/59; CH 310 formerly designated CHV/1/60.

The number immediately following any CV ("coryzavirus") designation in this table represents a type number assigned to the indicated strain by the investigators originally describing that strain (18, 26). "Coryzavirus" was the term originally used to describe these strains (15, 18, 26).

1. GWALTNEY, J. M., Jr., and W. S. JORDAN, Jr.: Bact. Rev. **28**, 409 (1964).
2. CHANOCK, R. M., M. A. MUFSON, and K. M. JOHNSON: Progr. med. Virol. **7**, 208 (1965).
3. PHILLIPS, C. A., S. RIGGS, J. L. MELNICK, and C. A. GRIM: J. Amer. med. Ass. **192**, 277 (1965).
4. MUFSON, M. A., P. A. WEBB, H. KENNEDY, V. GILL, and R. M. CHANOCK: J. Amer. med. Ass. **195,** 1 (1965).
5. TYRRELL, D. A. J.: in Common Colds and Related Diseases, 155 (The Williams and Wilkins Co., Baltimore, and Arnolds, London, 1965).
6. HAMRE, D., A. P. CONNELLY, Jr., and J. J. PROCKNOW: Amer. J. Epidem. **83**, 283 (1966).
7. GWALTNEY, J. M., and W. S. JORDAN, Jr.: Amer. Rev. resp. Dis. **43**, 362 (1966).
8. TYRRELL, D. A. J., and R. M. CHANOCK: Science **141**, 152 (1963).
9. KAPIKIAN, A. Z., M. A. MUFSON, H. D. JAMES, Jr., A. R. KALICA, H. H. BLOOM, and R. M. CHANOCK: Proc. Soc. exp. Biol. (N.Y.) **122**, 1155 (1966).
10. HAMPARIAN, V. V., and R. M. CONANT (in preparation).
11. PELON, W., W. J. MOGABGAB, I. A. PHILLIPS, and W. E. PIERCE: Bact. Proc. **67** (1956).
12. PRICE, W. H.: Proc. nat. Acad. Sci. (Wash.) **42**, 892 (1956).
13. TAYLOR-ROBINSON, D., and D. A. J. TYRRELL: Lancet **i**, 452 (1962).
14. MOGABGAB, W. J.: Amer. J. Hyg. **76**, 160 (1962).
15. HAMPARIAN, V. V., A. KETLER, and M. R. HILLEMAN: Proc. Soc. exp. Biol. (N.Y.) **108**, 444 (1961).
16. JOHNSON, K. M., H. H. BLOOM, R. M. CHANOCK, M. A. MUFSON, and V. KNIGHT: Amer. J. publ. Hlth **52**, 933 (1962).
17. JOHNSON, K. M., and L. ROSEN: Amer. J. Hyg. **77,** 15 (1963).
18. KETLER, A., V. V. HAMPARIAN, and M. R. HILLEMAN: Proc. Soc. exp. Biol. (N.Y.) **110**, 821 (1962).
19. HAMRE, D., A. P. CONNELLY, Jr., and J. J. PROCKNOW: J. Lab. clin. Med. **64,** 450 (1964).
20. Submitted by Dr. WILLIAM J. MOGABGAB, Tulane University, New Orleans.
21. CONNELLY, A. P., Jr., and D. HAMRE: J. Lab. Clin. Med. **63**, 30 (1964).
22. WEBB, P. A., K. M. JOHNSON, and M. A. MUFSON: Proc. Soc. exp. Biol. (N.Y.) **116**, 845 (1964).

(Continued Table 12)

23. MUFSON, M. A., R. KAWANA, H. D. JAMES, Jr., L. W. GAULD, H. H. BLOOM, and R. M. CHANOCK: Amer. J. Epidem. **81**, 32 (1965).
24. DICK, E. C., and C. R. BLUMER (in preparation).
25. PHILLIPS, C. A., J. L. MELNICK, and C. A. GRIM: Proc. Soc. exp. Biol. (N.Y.) **119**, 798 (1965).
26. HAMPARIAN, V. V., M. B. LEAGUS, and M. R. HILLEMAN: Proc. Soc. exp. Biol. (N.Y.) **116**, 976 (1964).
27. Submitted by Dr. E. H. LENNETTE and Dr. J. H. SCHIEBLE, California State Department of Public Health, Berkeley, California.
28. MELNICK, J. L. (personal communication).

disease shows that in the field and experimentally new serotypes of a picorna-virus may be evolved if conditions are appropriate. Prolonged epidemiological studies will be necessary to give a final answer to the question.

So far, few serotypes of rhinoviruses of animals have been described, although the syndrome of cat influenza, a common disease, especially of young animals, seems to be due to several viruses which are antigenically distinct; several of them are poised on the borderline between enteroviruses and typical rhinoviruses but others may not be rhinoviruses at all (BÜRKI, 1965; ZWILLENBERG and BÜRKI, 1966). Respiratory disease due to equine rhinoviruses has been described so far in Britain, Canada, and South Africa (PLUMMER, 1963; DITCHFIELD and MACPHER-SON, 1965; ERASMUS, 1966, personal communication). The one serotype of rhino-virus isolated from a calf was isolated from a mild illness and caused little effect when administered experimentally (BÖGEL and BÖHM, 1962)[3]. We have no in-formation about how bovine rhinoviruses spread and persist under normal conditions of husbandry. Recent work on Foot-and-Mouse disease virus suggests that bovine carriers might exist. There is no evidence as yet that these viruses are transmissible from or to man in nature. The calf rhinovirus has no effect on human cells *in vitro*, but some human rhinoviruses will produce degeneration in calf kidney cells (VARGOSKO, personal communication), although they have not been adapted to them. The cat rhinoviruses do not affect human cells *in vitro* but the equine rhinovirus was given in a very large dose to a volunteer and he became infected and developed pharyngitis; in addition, stable workers carry antibody against the virus (PLUMMER, 1963).

IV. Appendix

Diagnostic Tests for Rhinovirus Infection

1. Clinical Specimens

The following have been successfully used:

a) Nasal washings taken with saline and supplemented with an equal volume of bacteriological broth or a final concentration of 0.2% to 2% bovine plasma albumin.

b) Nasal swabs (it is important that the swab is well wetted with nasal secretions), nasal "blow-outs" into a Petri dish. These are placed in a bottle with

---

[3] Further isolations have now been reported (WIZIGMANN and SCHIEFFER, 1966).

transport medium, such as Hanks' saline containing 2% bovine plasma albumin, or veal infusion broth.

c) Throat swabs collected like nasal swabs — these are often satisfactory, but may contain less virus than nasal swabs.

2. Paired Sera

The first serum should be collected in the acute phase of the illness. The time of development of antibody is not precisely known but it may well be rather slow so a second specimen should not be taken before two to three weeks after the illness.

3. Laboratory Tests

The specimens should if possible be inoculated immediately into cultures but there is little disadvantage in freezing them at $-70°C$ or lower. Cultures of human embryo cells are prepared, either a sensitive strain of human diploid fibroblasts such as WI 38 or primary or secondary human-embryo-kidney-cells. The former are maintained in 2% calf serum in 98% Eagle's medium, the latter cells are maintained in 2% calf serum and 0.25% lactalbumin hydrolysate in Hanks' saline buffered to give a pH of between about 7.3 and 7.0. Mixtures of Eagle's medium and medium 199 have been used successfully provided that the proper pH was obtained and the cultures were changed as necessary to prevent it falling too much. Even a mild degree of "toxicity" of the inoculum may interfere with virus isolation, and it is therefore often helpful to change the medium on the day after cultures have been inoculated. The amount of virus in a specimen may be quite small, therefore it is better to use a large total volume of inoculum. We add 0.2 ml of medium to each of three tubes. Cultures are always incubated in a roller drum at 33°C. If possible organ cultures of human embryo trachea or nose should also be inoculated, using methods already described (HOORN, 1966; TYRRELL and BLAMIRE, 1967; HIGGINS, 1966 b). After 1 or 2 passages in these cultures the medium should be inoculated into tissue cultures as just described.

The cultures should be inspected regularly, if possible daily or on alternate days, because a focal cytopathic effect in human kidney cells may disappear a day or two after it appears, and passage of rhinoviruses may be difficult if not attempted when the cytopathic effect is steadily progressing. Passage is most likely to be successful if the medium is transferred to new cultures as soon as it seems that the number of freshly affected cells in the culture has reached a maximum. It is better not to store culture media between passages.

A presumptive diagnosis of a rhinovirus infection can be made if a cytopathic effect is seen which resembles that of picornaviruses and develops rather slowly. However, the diagnosis must be confirmed. This may be done by comparing the cytopathic effect in further cultures, one set of which are set up as in the virus isolation while the others are maintained at pH 7.6 at 37°C in a stationary rack. Rhinoviruses fail to grow under the latter conditions. This is however a less decisive test than that of showing that the virus is acid-labile.

This may be done by mixing one volume of virus with one volume of $0.1 M$ phosphate buffer at pH 5 or $0.1 M$ sodium citrate citric acid buffer at pH 4. The mixtures are held at 37°C for one hour and then neutralized by the addition

Table 13. *The Isolation of Certain Serotypes of Rhinoviruses as*

	1A	1B	2	3	4	5	6	7	8	9	10	11	12	13	14	15	16	17	18
1959—1960 Pennsylvania and New Jersey, U.S.A. (1)								9		1	4	1	3						3
1960—1961 Pennsylvania and New Jersey, U.S.A. (1)		1		1					8							3	3		
1961 N. Carolina, U.S.A. (2)	24	4	18				1		1			1	1			3			
1961—1962 California, U.S.A.		9	13				1		1								1		
1962 United Kingdom		2			1											1			
1962—1963 California, U.S.A.	1	3			5														
1963 United Kingdom	1	6	2		2		1		1				1						
Virginia, U.S.A.		1			7							2			4	1		2	
Texas, U.S.A.																1			
1963—1964 California, U.S.A.	4	1		1	1				20	3			1			1			
1964 United Kingdom	3	4																	
1964—1965 Virginia, U.S.A.	1	1		1	2		6	8		2		1		1	1			4	
California, U.S.A.	3	6	4					1							1				1
1965 United Kingdom	5	3	7		3								1		2	1			
Virginia, U.S.A.			1		1		1			1		3			12				
1965—1966 California, U.S.A.	15		26							1	1				1		1		3
1966 United Kingdom	5		4		2	1	1				1								

These results should be viewed in conjunction with those of HAMRE and PROCKNOW shown in
Certain rhinovirus serotypes have also been identified in other areas, e.g. Prague-types 2, 30 and
I am extremely grateful to the above and to the following workers who provided unpublished
Dr. MARGUERITE S. PEREIRA and Dr. MARY ROWLEY, Central Public Health Laboratory, Colindale
Dr. J. M. GWALTNEY, Dr. W. S. JORDAN, Jr., Dr. J. O. HENDLEY and Dr. G. SIMON, Charlottesville,
Dr. J. SCHIEBLE and Dr. E. D. LENNETTE, Department of Public Health, Berkeley, California,
(1) HAMPARIAN et al. (1964b). (2) MUFSON et al. (1966).

of one volume of 0.5 M buffer pH 7.2. This mixture diluted 1:5 is then inoculated
in tissue cultures along with a parallel culture which is incubated at pH 7 with
0.1 M phosphate buffer before addition of 0.5 M buffer. Cultures should be changed
about three hours after inoculation.

    One alternative method is to mix one portion of the virus with nine volumes
of Eagle's medium without bicarbonate (final pH 3.0) and another portion

Part of Systematic Surveys of Acute Respiratory Tract Infections

19	20	21	22	23	24	25	26	27	28	29	30	31	32	33	34	35	36	37	38	39	40	41	42	43	44	45	46	47
3	1	1	1	1	1	1	1	1	1	1																		
			1					1																				
						1				11	25	21	1			2					1	3	1	18	3			
						1				2		1	1		1													
										1																		
		1	1	1	1	4					4															6		
										3			1															
						3			1	5			4	1		2			1	13		1					6	8
										1		1			4												5	3
1				10						12	2	3														15		
													1															
		3	1	1	4				2	1		3				2							2				1	
4	1			10	2						1	1		6		2			1									
	4	1	1	5						4		1		3	1	1	9		1				2				1	
1				6						1	8	1			1							1						
										2	1		1															

Fig. 14.

31 by Dr. STRIZOVA, Glasgow — types 1B, 3, 4, 7, 10, 11, 12, 14, 16, 18, 29 and 32 by Dr. E. J. STOTT.
results which have been incorporated in this table.
Avenue, London, N.W. 9, England.
Virginia, U.S.A.
U.S.A.

with nine volumes of Eagle's medium neutralized with 0.01 M Tris buffer. The mixtures are held at room temperature for three hours before titration.

The most concentrated virus tested after treatment with acid should be noninfectious and the control sample should cause a cytopathic effect when diluted at least ten-fold more than this.

It is often desirable to test for ether-stability also. Virus is mixed with 20%

of anaesthetic ether and held overnight at 4°C alongside an untreated control sample. Next morning the ether is removed by standing the plugged tube in a water bath, or by exposing to a stream of sterile air or nitrogen. The control is similarly treated. The surviving virus is titrated and the titre should be within one ten-fold dilution of that of the control sample.

Virus size may be roughly determined by filtration through gradocol or millipore membranes of about 50 mμ APD. Virus is clarified by centrifugation at about 500 g for 30 min. or by passage through a filter of APD about 3–500 mμ. The filter is then set up and "satisfied" by passing through it 5% horse serum in saline or culture medium. We use about 10 ml for a 40 mm membrane. The virus is then passed through the filter and the filtrate and starting fluid are titrated. Up to 100-fold reductions in titre may occur with certain viruses even if the filter has been carefully "satisfied" and the fluid has been clarified. Therefore a negative result, *i.e.* failure to pass the membrane, is only satisfactory if the starting fluid has a titre of at least 10^2 ID$_{50}$.

Acknowledgements

I wish to thank Mrs. E. V. Sparrow and the Photographic Department of the National Institute for Medical Research for help in preparation of the typescript and illustrations.

References[4]

Almeida, J. D., and D. A. J. Tyrrell: The morphology of three previously uncharacterized human respiratory viruses that grow in organ culture. J. gen. Virol. **1**, 175–178 (1967).

Anderson, T. O., L. J. M. Riff, and G. G. Jackson: Immunoelectrophoresis of nasal secretions collected during a common cold: observations which suggest a mechanism of seroimmunity in viral respiratory infections. J. Immunol. **89**, 691–697 (1962).

Andrewes, C. H.: The common cold. Scientific American, **184**, 39–45 (1951).

Andrewes, C. H.: The common cold. Brit. med. Bull. **9**, 206–207 (1953).

Andrewes, C. H.: The taxonomic position of common cold viruses and some others. Yale J. Biol. Med. **34**, 200–206 (1961–1962).

Andrewes, C. H.: The complex epidemiology of respiratory virus infections. Science **146**, 1274–1277 (1964).

Andrewes, C. H., D. M. Chaproniere, A. E. H. Gompels, H. G. Pereira, and A. T. Roden: Propagation of common-cold virus in tissue cultures. Lancet **2**, 546–547 (1953).

Andrewes, C. H., and W. G. Oakley: St. Bart. Hosp. Rep. **40**, 74 (1933).

Armstrong, J. A., and D. A. J. Tyrrell: (1963). Unpublished work quoted in Tyrrell (1963a).

Behbehani, A. M., J. L. Melnick, and M. E. DeBakey: Continuous cell strains derived from human atheromatous lesions and their viral susceptibility. Proc. Soc. exp. Biol. (N.Y.) **118**, 759–763 (1965).

Bloom, H. H., B. R. Forsyth, K. M. Johnson, and R. M. Chanock: Relationship of rhinovirus infection to mild upper respiratory disease. 1. Results of a survey in young adults and children. J. Amer. med. Ass. **186**, 38–45 (1963).

Bögel, K., und H. Böhm: Ein Rhinovirus des Rindes. Zbl. Bakt. I. Abt. Orig. **187**, 2–14 (1962).

Bögel, K., und A. Mayr: Ein Resistenztest mit Detergentien zur Charakterisierung von Viren. Zbl. Bakt. I. Abt. Orig. **186**, 134–138 (1962).

---

[4] The references are believed to be complete up to 1966, although some review articles and some papers of a preliminary nature have been omitted. The survey of the literature included a MEDLARS search performed in the autumn of 1966. A review stressing the epidemiology of rhinoviruses has been published by Dr. D. Hamre as Vol. 1 of Monographs in Virology, Karger, Basel, in 1968.

BROWN, P. K., and D. TAYLOR-ROBINSON: Respiratory virus antibodies in sera of persons living in isolated communities. Bull. Wld Hlth Org. **34**, 895—900 (1966).

BROWN, P. K., and D. A. J. TYRRELL: Experiments on the sensitivity of strains of human fibroblasts to infection with rhinoviruses. Brit. J. exp. Path. **45**, 571—578 (1964).

BROWN, P. K., D. A. J. TYRRELL, and J. P. JACOBS: Sensitivity of human diploid cell strains (HDCS) to rhinoviruses. Proc. Symposium on the characterization and uses of human diploid cell strains. Opatija 251—254 (1963).

BRYAN, W. T. K., and M. P. BRYAN: Structural changes in the ciliated epithelial cells during the common cold. Trans. Amer. Acad. Ophthal. Otolaryng. **57**, 297—303 (1953).

BUCKLAND, F. E., M. L. BYNOE, and D. A. J. TYRRELL: Experiments on the spread of colds. II. Studies in volunteers with coxsackie virus A 21. J. Hyg. (Lond.) **63**, 327—343 (1965).

BUCKLAND, F. E., J. E. DOGGETT, and D. A. J. TYRRELL: The specificity of antibody responses of human volunteers to certain respiratory viruses. J. Hyg. (Lond.) **62**, 115—119 (1964).

BUCKLAND, F. E., and D. A. J. TYRRELL: Loss of infectivity on drying various viruses. Nature (Lond.) **195**, 1063—1064 (1962).

BÜRKI, F.: Picornaviruses of cats. Arch. ges. Virusforsch. **15**, 690—696 (1965).

BYNOE, M. L., D. HOBSON, J. HORNER, A. KIPPS, G. C. SCHILD, and D. A. J. TYRRELL: Inoculation of human volunteers with a strain of virus isolated from a common cold. Lancet **1**, 1194—1196 (1961).

CATE, T. R., R. B. COUCH, W. F. FLEET, W. R. GRIFFITH, P. J. GERONE, and V. KNIGHT: Production of tracheo bronchitis in volunteers with rhinovirus in a small-particle aerosol. Amer. J. Epidem. **81**, 95—105 (1965).

CATE, T. R., R. B. COUCH, and K. M. JOHNSON: Studies with rhinoviruses in volunteers: production of illness, effect of naturally acquired antibody, and demonstration of a protective effect not associated with serum antibody. J. clin. Invest. **43**, 56—67 (1964).

CHAPPLE, P. J., and J. HARRIS: Biophysical studies of a rhinovirus. Ultracentrifugation and electron microscopy. Nature (Lond.) **209**, 790—792 (1966).

CHAPPLE, P. J., B. RIDGEWELL, and D. A. J. TYRRELL: A complement fixing antigen from an M rhinovirus. Arch. ges. Virusforsch. **21**, 123—126 (1967).

CONNELLY, A. P., Jr., and D. HAMRE: Virologic studies on acute respiratory disease in young adults. II. Characteristics and serological studies of three new rhinoviruses. J. Lab. clin. Med. **63**, 30—43 (1964).

COUCH, R. B., T. R. CATE, G. DOUGLAS, Jr., P. J. GERONE, and V. KNIGHT: Effect of route of inoculation on experimental respiratory viral disease in volunteers and evidence for airborne transmission. Bact. Rev. **30**, 517—529 (1966).

COUCH, R. B., T. R. CATE, P. J. GERONE, W. F. FLEET, D. J. LANG, W. R. GRIFFITH, and V. KNIGHT: Production of illness with a small-particle aerosol of coxsackie virus A 21. J. clin. Invest. **44**, 535—542.

DANS, P. E., B. R. FORSYTH, and R. M. CHANOCK: Density of infectious virus and complement-fixing antigens of two rhinovirus strains. J. Bact. **91**, 1605—1611 (1966).

DEIBEL, R., and C. P. DUCHARME: Recovery of strain of rhinovirus from infant with upper respiratory infection. N.Y. St. J. Med. **65**, 412—416 (1965).

DIMMOCK, N. J.: Biophysical studies of a rhinovirus. Extraction and assay of infectious ribonucleic acid. Nature (Lond.) **209**, 790—794 (1966).

DIMMOCK, N. J.: Differences between the thermal inactivation of picornaviruses at "high" and "low" temperatures. Virology **31**, 338—353 (1967).

DIMMOCK, N. J.: (1968), in preparation.

DIMMOCK, N. J., and D. A. J. TYRRELL: Physicochemical properties of some viruses isolated from common colds (rhinoviruses). Lancet **2**, 536—537 (1962).

DIMMOCK, N. J., and D. A. J. TYRRELL: Some physicochemical properties of rhino-viruses. Brit. J. exp. Path. **45**, 271—280 (1964).

DITCHFIELD, J., and L. E. MACPHERSON: The properties and classification of two new rhinoviruses recovered from horses in Toronto, Canada. Cornell Vet. **55**, 182—189 (1965).

DOCHEZ, A. R., K. C. MILLS, and Y. KNEELAND: Filterable viruses in infection of the upper respiratory tract. J. Amer. med. Ass. **110**, 177—180 (1938).

DOCHEZ, A. R., G. S. SHIBLEY, and K. C. MILLS: Studies in the common cold. IV. Experimental transmission of the common cold to anthropoid apes and human beings by means of a filterable agent. J. exp. Med. **52**, 701—716 (1930).

DOGGETT, J. E., M. L. BYNOE, and D. A. J. TYRRELL: Some attempts to produce an experimental vaccine with rhinoviruses. Brit. med. J. **1**, 34—36 (1963).

DOUGLAS, R. G., Jr., R. H. ALFORD, T. R. CATE, and R. B. COUCH: The leukocyte response during viral respiratory illness in man. Ann. intern. Med. **64**, 521—530 (1966a).

DOUGLAS, R. G., Jr., T. R. CATE, and R. B. COUCH: Growth and cytopathic effect of H type rhinoviruses in monkey kidney tissue culture. Proc. Soc. exp. Biol. (N.Y.) **123**, 238—241 (1966b).

DOUGLAS, R. G., Jr., T. R. CATE, P. J. GERONE, and R. B. COUCH: Quantitative rhinovirus shedding patterns in volunteers. Amer. Rev. resp. Dis. **94**, 159—167 (1966c).

DREIZIN, R. S., V. D. BELYATSKY, M. P. CHUMAKOV, E. S. KETILADZE, M. E. SUK-HAREVA, L. A. KONSTANTINOVA, A. N. MUSTAFINA, I. A. LUTSEVICH, K. E. SMA-GULOV, and S. A. VEISERIK: (In Russian.) On the spread of rhinovirus infection among adults and young children. Vop. Virus. (2) 214—220 (1966).

DRESCHER, J., and K. SCHRADER: Titration of poliovirus and influenza virus by means of the haemaggregation test. Amer. J. Hyg. **79**, 218—235 (1964).

EADIE, M. B., E. J. STOTT, and N. R. GRIST: Virological studies on bronchitis. Brit. med. J. **2**, 671—673 (1966).

EGGERS, H. J., and I. TAMM: 2-(Hydroxybenzyl)-benzimidazole (HBB) as an aid in virus classification. Virology **13**, 545—546 (1961).

ELVEBACK, L. R., J. P. FOX, A. KETLER, C. D. BRANDT, F. E. WASSERMANN, and C. E. HALL: The virus watch program: a continuing surveillance of viral infections in metropolitan New York families. III. Preliminary report on association of infections with disease. Amer. J. Epidem. **88**, 436—454.

FENTERS, J. D., S. S. GILLUM, J. C. HOLPER, and G. S. MARQUIS: Serotypic relationships among rhinoviruses. Amer. J. Epidem. **84**, 10—20 (1966).

FIALA, M., and G. E. KENNY: Enhancement of rhinovirus plaque formation in human heteroploid cell cultures by magnesium and calcium. J. Bact. **92**, 1710—1715 (1966).

FLEET, W., R. B. COUCH, T. R. CATE, and V. KNIGHT: Homologous and heterologous resistance to rhinovirus common cold. Amer. J. Epidem. **82**, 185—196 (1965).

FORSYTH, B. R., H. H. BLOOM, K. M. JOHNSON, and R. M. CHANOCK: Patterns of illness in rhinovirus infections of military personnel. New Engl. J. Med. **269**, 602—606 (1963).

FOSTER, G. B.: The etiology of common colds. J. Amer. med. Ass. **66**, 1180—1182 (1916).

GWALTNEY, J. M.: Micro-neutralization test for identification of rhinovirus sero-types. Proc. Soc. exp. Biol. (N.Y.) **122**, 1137—1141 (1966).

GWALTNEY, J. M., J. O. HENDLEY, G. SIMON, and W. S. JORDAN, Jr.: Rhinovirus infections in an industrial population. 1. The occurrence of illness. New Engl. J. Med. **275**, 1261—1268 (1966).

GWALTNEY, J. M., Jr., and W. S. JORDAN, Jr.: Rhinoviruses and respiratory disease. Bact. Rev. **28**, 409—422 (1964).

HAFF, R. F., B. WOHLSEN, E. E. FORCE, and R. C. STEWART: Growth characteristics of two rhinovirus strains in WI-26 and monkey kidney cells. J. Bact. **91**, 2339—2342 (1966).

HAMPARIAN, V. V., M. R. HILLEMAN, and A. KETLER: Contributions to characterization of animal viruses. Proc. Soc. exp. Biol. (N.Y.) 112, 1040—1050 (1963).

HAMPARIAN, V. V., A. KETLER, and M. R. HILLEMAN: Recovery of new viruses (coryzavirus) from cases of common cold in human adults. Proc. Soc. exp. Biol. (N.Y.) 108, 444—453 (1961).

HAMPARIAN, V. V., A. KETLER, and M. R. HILLEMAN: The ECHO 28 rhinovirus-coryzavirus (ERC) group of viruses. Amer. Rev. resp. Dis. 88, part 2, 269—273 (1963).

HAMPARIAN, V. V., M. B. LEAGUS, and M. R. HILLEMAN: Additional rhinovirus serotypes. Proc. Soc. exp. Biol. (N.Y.) 116, 976—984 (1964a).

HAMPARIAN, V. V., M. B. LEAGUS, M. R. HILLEMAN, and J. STOKES, Jr.: Epidemiological investigations of rhinovirus infections. Proc. Soc. exp. Biol. (N.Y.) 117, 469—476 (1964b).

HAMRE, D., A. P. CONNELLY, and J. J. PROCKNOW: Virologic studies of acute respiratory disease in young adults. III. Some biologic and serologic characteristics of seventeen rhinovirus serotypes isolated October, 1960 to June, 1961. J. Lab. clin. Med. 64, 450—460 (1964).

HAMRE, D., A. P. CONNELLY, and J. J. PROCKNOW: Virologic studies of acute respiratory disease in young adults. IV. Virus isolations during four years of surveillance. Amer. J. Epidem. 83, 233—249 (1966).

HAMRE, D., and J. J. PROCKNOW: Virological studies in acute respiratory disease in young adults. Proc. Soc. exp. Biol. (N.Y.) 107, 770—773 (1961a).

HAMRE, D., and J. J. PROCKNOW: Viruses isolated from natural common colds in the U.S.A. Brit. med. J. 2, 1382—1385 (1961b).

HAMRE, D., and J. J. PROCKNOW: Virological studies on common colds among young adult medical students. Amer. Rev. resp. Dis. 88, 277—281 (1963).

HAYFLICK, L., and P. S. MOORHEAD: The serial cultivation of human diploid cell strains. Exp. Cell Res. 25, 585—621 (1961).

HEMMES, J. H., K. C. WINKLER, and S. M. KOOL: Virus survival as a seasonal factor in influenza and poliomyelitis. Nature (Lond.) 188, 430—431 (1960).

HENIGST, W.: Nicht identifizierbare fragliche Enteroviren in einem Kinderheim. Zbl. Bakt. I. Abt. Orig. 197, 1—12 (1965).

HIGGINS, P. G.: The isolation of viruses from acute respiratory infections. Part IV: A comparative study of the use of cultures of human embryo kidney and human embryo diploid fibroblasts (WI 38). Mth. Bull. Minist. Hlth Lab. Serv. 25, 223—229 (1966a).

HIGGINS, P. G.: The isolation of viruses from acute respiratory infections. Part V: The use of organ cultures of human embryonic nasal and tracheal ciliated epithelium. Mth. Bull. Minist. Hlth Lab. Serv. 25, 283—288 (1966b).

HIGGINS, P. G., D. G. BOSTON, and E. M. ELLIS: The isolation of viruses from acute respiratory infections. II. A study of the isolations made from cases occurring in a general practice in 1963. Mth. Bull. Minist. Hlth Lab. Serv. 23, 93—105 (1964).

HIGGINS, P. G., E. M. ELLIS, and D. G. BOSTON: The isolation of viruses from acute respiratory infections. A study of the isolations made from cases occurring in a general practice in 1962. Mth. Bull. Minist. Hlth Lab. Serv. 22, 71—84 (1963).

HILDING, A. C.: The common cold. Arch. Otolaryng. 12, 133—150 (1930).

HILLEMAN, M. R., C. M. REILLY, J. STOKES, Jr., and V. V. HAMPARIAN: Clinical-epidemiological findings in coryzavirus infections. Amer. Rev. resp. Dis. 88, 274—276 (1963).

HITCHCOCK, G., and D. A. J. TYRRELL: Some virus isolations from common colds. II. Virus interference in tissue cultures. Lancet 1, 237—239 (1960).

HOBSON, D., and G. C. SCHILD: Virological studies in natural common colds in Sheffield in 1960. Brit. med. J. 2, 1414—1418 (1960).

HOLZEL, A., L. PARKER, W. H. PATTERSON, D. CARTMEL, L. L. R. WHITE, R. PURDY, K. M. THOMPSON, and J. O'H. TOBIN: Virus isolations from throats of children admitted to hospital with respiratory and other diseases, Manchester 1962—1964. Brit. med. J. 1, 614—619 (1965a).

HOLZEL, A., P. A. SMITH, and J. O'H. TOBIN: A new type of meningo-encephalitis associated with a rhinovirus. Acta paediat. (Uppsala) **54**, 168—174 (1965b).

HOORN, B.: Organ cultures of ciliated epithelium for the study of respiratory viruses. Acta path. microbiol. scand. **66**, Suppl. 183 (1966).

HOORN, B., M. L. BYNOE, P. J. CHAPPLE, and D. A. J. TYRRELL: Inoculation of a novel type of rhinovirus (HS) to human volunteers. Arch. ges. Virusforsch. **18**, 226—230 (1966).

HOORN, B., and D. A. J. TYRRELL: On the growth of certain "newer" respiratory viruses in organ culture. Brit. J. exp. Path. **46**, 109—118 (1965).

HOORN, B., and D. A. J. TYRRELL: A new virus cultivated only in organ cultures of human ciliated epithelium. Arch. ges. Virusforsch. **18**, 210—225 (1966).

HOPE-SIMPSON, R. E.: A long-term study of acute respiratory diseases in a natural community. Proc. roy. Soc. Med. **59**, 639—642 (1966).

International Enterovirus Study Group: Picornavirus Group. Virology **19**, 114—116 (1963).

JACKSON, G. G., and H. F. DOWLING: Transmission of the common cold to volunteers under controlled conditions. IV. Specific immunity to the common cold. J. clin. Invest. **38**, 762—769 (1959).

JACKSON, G. G., H. F. DOWLING, L. W. AKERS, R. L. MULDOON, A. V. DYKE, and G. C. JOHNSON: Immunity to the common cold from protective serum antibody. Time of appearance, persistence and relation to re-infection. New Engl. J. Med. **266**, 791—796 (1962).

JACKSON, G. G., H. F. DOWLING, and T. O. ANDERSON: Neutralization of common cold agents in volunteers by pooled gamma globulin. Science **128**, 27—28 (1958).

JACKSON, G. G., H. F. DOWLING, J. O. ANDERSON, L. RIFF, J. SAPORTA, and M. TURCK: Susceptibility and immunity to common upper respiratory viral infections — the common cold. Ann. intern. Med. **53**, 719—738 (1960b).

JACKSON, G. G., H. F. DOWLING, and W. J. MOGABGAB: Infectivity and interrelationships of 2060 and JH viruses in volunteers. J. Lab. clin. Med. **55**, 331—341 (1960a).

JACKSON, G. G., H. F. DOWLING, I. G. SPIESMAN, and A. V. BOAND: Transmission of the common cold to the volunteers under controlled conditions. Arch. intern. Med. **101**, 267—278 (1958).

JOHNSON, K. M., H. H. BLOOM, R. M. CHANOCK, M. A. MUFSON, and V. KNIGHT: VI. The newer enteroviruses. Amer. J. publ. Hlth **52**, 933—940 (1962).

JOHNSON, K. M., H. H. BLOOM, B. R. FORSYTH, and R. M. CHANOCK: Relationship of rhinovirus infection to mild upper respiratory disease. II. Epidemiological observations in male military trainees. Amer. J. Epidem. **81**, 131—139 (1965).

JOHNSON, K. M., H. H. BLOOM, M. A. MUFSON, and R. M. CHANOCK: Acute respiratory disease associated with Coxsackie A 21 virus infection. J. Amer. med. Ass. **179**, 112—119 (1962).

JOHNSON, K. M., and L. ROSEN: Characteristics of five newly recognised enteroviruses recovered from the human oropharynx. Amer. J. Hyg. **77**, 15—25 (1963).

KAPIKIAN, A. Z., and 17 others: A rhinovirus numbering system. Nature (Lond.) **213**, 761—763 (1967).

KAPIKIAN, A. Z., M. A. MUFSON, H. D. JAMES, Jr., A. R. KALICA, H. H. BLOOM, and R. M. CHANOCK: Characterization of two newly recognized rhinovirus serotypes of human origin. Proc. Soc. exp. Biol. (N.Y.) **122**, 1155—1162 (1966).

KASEL, J., and V. KNIGHT: Infection of human volunteers with Pett virus. Proc. Soc. exp. Biol. (N.Y.) **113**, 602—605 (1963).

KAWANA, R., R. M. CHANOCK, and M. A. MUFSON: Unpublished data quoted in MUFSON et al. (1965).

KAWANA, R., S. YOSHIDA, I. MATSUMOTO, M. KANEKO, H. WAKO, and S. YASUOKA: Rhinoviruses isolated from Japanese children with common cold. Jap. J. Microbiol. **10**, 127—128 (1966).

KENDALL, E. J., M. L. BYNOE, and D. A. J. TYRRELL: Virus isolations from common colds occurring in a residential school. Brit. med. J. **2**, 82—86 (1962).

KETLER, A., V. V. HAMPARIAN, and M. R. HILLEMAN: Characterization and classification of ECHO 28-rhinovirus-coryzavirus agents. Proc. Soc. exp. Biol. (N.Y.) **110**, 821—831 (1962).

KNIGHT, V.: The use of volunteers in medical virology. Progr. med. Virol. **6**, 1—26 (1964).

KRUSE, W. v.: Die Erreger von Husten und Schnupfen. Münch. med. Wschr. **61**, 1547 (1914).

LIDWELL, O. M., R. W. MORGAN, and R. E. O. WILLIAMS: The epidemiology of the common cold. IV. The effect of weather. J. Hyg. (Lond.) **63**, 427—439 (1965).

MCGREGOR, S., C. A. PHILLIPS, and H. D. MAYOR: Purification and biophysical properties of rhinoviruses. Proc. Soc. exp. Biol. (N.Y.) **122**, 118—121 (1966).

MASCOLI, C. C., M. B. LEAGUS, R. E. WEIBEL, J. STOKES, Jr., H. REINHART, and M. R. HILLEMAN: Attempt at immunization by oral feeding of live rhinoviruses in enteric-coated capsules. Proc. Soc. exp. Biol. (N.Y.) **121**, 1264—1268 (1966).

MAYOR, H. D.: Picornavirus symmetry. Virology **22**, 156—160 (1964).

MOGABGAB, W. J.: 2060 virus (ECHO 28) in KB cell cultures. Characteristics, complement fixation and relationship to some other respiroviruses. Amer. J. Hyg. **76**, 15—26 (1962).

MOGABGAB, W. J.: Additional respirovirus type related to GL 2060 (ECHO 28) virus, from military personnel, 1959. Amer. J. Hyg. **76**, 160—172 (1963a).

MOGABGAB, W. J.: Upper respiratory illness vaccines. Perspectives and trials. Ann. intern. Med. **57**, 526—537 (1963b).

MOGABGAB, W. J., and B. HOLMES: 2060 and JH viruses in secondary monkey kidney cultures. J. infect. Dis. **108**, 59—62 (1961).

MOGABGAB, W. J., and W. PELON: Problems in characterizing and identifying an apparently new virus found in association with mild respiratory disease in recruits. Ann. N.Y. Acad. Sci. **67**, 403—412 (1957).

MONTO, A. S. and K. M. JOHNSON: Serologic relationships of the B 632 and ECHO-28-rhinovirus strains. Proc. Soc. exp. Biol. (N.Y.) **121**, 615—619 (1966).

MUFSON, M. A., H. H. BLOOM, B. R. FORSYTH, and R. M. CHANOCK: Relationship of rhinovirus infection to mild upper respiratory disease. III. Further epidemiological observations in military personnel. Amer. J. Epidem. **83**, 379—388 (1966a).

MUFSON, M. A., R. KAWANA, H. D. JAMES, Jr., L. W. GAULD, H. H. BLOOM, and R. M. CHANOCK: A description of six new rhinoviruses of human origin. Amer. J. Epidem. **81**, 32—43 (1965).

MUFSON, M. A., W. M. LUDWIG, H. D. JAMES, L. W. GAULD, J. A. ROURKE, J. C. HOLPER, and R. M. CHANOCK: Effect of neutralizing antibody on experimental rhinovirus infection. J. Amer. med. Ass. **186**, 578—584 (1963).

MUFSON, M. A., P. A. WEBB, H. KENNEDY, V. GILL, and R. M. CHANOCK: Etiology of upper respiratory tract illnesses among civilian adults. J. Amer. med. Ass. **195**, 1—7 (1966b).

PARSONS, R., and D. A. J. TYRRELL: A plaque method for assaying some viruses isolated from common colds. Nature (Lond.) **189**, 640—642 (1961).

PARSONS, R., and D. A. J. TYRRELL: (1962) unpublished.

PELON, W.: Classification of the "2060" virus as ECHO 28 and further study of its properties. Amer. J. Hyg. **73**, 36—54 (1961).

PELON, W., and W. J. MOGABGAB: Further studies on 2060 virus. Proc. Soc. exp. Biol. (N.Y.) **102**, 392—395 (1959).

PELON, W., W. J. MOGABGAB, I. A. PHILLIPS, and W. E. PIERCE: A cytopathogenic agent isolated from normal recruits with mild respiratory illnesses. Proc. Soc. exp. Biol. (N.Y.) **94**, 262—267 (1957).

PEREIRA, M. S., M. H. HAMBLING, J. C. MCDONALD, and A. J. ZUCKERMAN: Viruses from the common cold. A survey in Royal Air Force recruits on arrival from civilian life. J. Hyg. (Lond.) **61**, 471—478 (1963).

PHILLIPS, C. A., J. L. MELNICK, and C. A. GRIM: Characterization of 3 new rhinovirus serotypes. Proc. Soc. exp. Biol. (N.Y.) **119**, 798—801 (1965a).

PHILLIPS, C. A., S. RIGGS, J. L. MELNICK, and C. A. GRIM: Rhinoviruses associated with common colds in a student population. J. Amer. med. Ass. **192**, 277—280 (1965 b).

PLUMMER, G.: An equine respiratory enterovirus: some biological and physical properties. Arch. ges. Virusforsch. **12**, 694—700 (1963).

PLUMMER, G.: The picornaviruses of man and animals: A comparative review. Progr. med. Virol. **7**, 326—361 (1965).

PORTERFIELD, J. S.: Titration of some common cold viruses (rhinoviruses) and their antisera by a plaque method. Nature (Lond.) **194**, 1044—1047 (1962).

PORTNOY, B., H. L. ECKERT, and M. A. SALVATORE: Rhinovirus infection in children with acute lower respiratory disease. Evidence against etiological importance. Pediatrics **35**, 899—905 (1965).

PRICE, W. H.: The isolation of a new virus associated with respiratory clinical disease in humans. Proc. nat. Acad. Sci. (Wash.) **42**, 892—896 (1956).

PRICE, W. H.: Vaccine for the prevention in humans of cold-like symptoms associated with the JH virus. Proc. nat. Acad. Sci. (Wash.) **43**, 790—795 (1957).

PRICE, W. H., H. EMERSON, I. IBLER, R. LACHAINE, and A. TERRELL: Studies of the JH and 2060 viruses and their relationship to mild upper respiratory disease in humans. Amer. J. Hyg. **69**, 224—249 (1959).

REILLY, C. M., S. M. HOCH, J. STOKES, Jr., L. McCLELLAND, V. V. HAMPARIAN, A. KETLER, and M. R. HILLEMAN: Clinical and laboratory findings in cases of respiratory illness caused by coryzaviruses. Ann. intern. Med. **57**, 515—525 (1962).

REMINGTON, J. S., K. L. VOSTI, A. LIETZE, and A. L. ZIMMERMAN: Serum proteins and antibody activity in human nasal secretions. J. clin. Invest. **43**, 1613—1624 (1964).

RITOVA, V. V., V. M. ZHDANOV, and E. I. SCHASTNY: Serological evidence of respiratory virus infections in Moscow children in 1948 to 1956. Pediatrics **38**, 402—404 (1966).

RODEN, A. T.: Clinical assessment of the common cold. Proc. roy. Soc. Med. **51**, 271—273 (1958).

ROSEN, L.: Subclassification of picornaviruses. Bact. Rev. **29**, 173—184 (1965).

ROSSEN, R. D., R. H. ALFORD, W. T. BUTLER, and W. E. VANMIER: The separation and characterization of proteins intrinsic to nasal secretion. J. Immunol. **97**, 369—378 (1966 a).

ROSSEN, R. D., R. G. DOUGLAS, Jr., T. R. CATE, R. B. COUCH, and W. T. BUTLER: The sedimentation behaviour of rhinovirus neutralizing activity in nasal secretion and serum following the rhinovirus common cold. J. Immunol. **97**, 532—538 (1966 b).

SCHILD, G. C., and D. HOBSON: Neutralizing antibody levels in human sera with the HGP and B632 strains of common cold virus. Brit. J. exp. Path. **43**, 288—294 (1962).

Scientific Committee on Interferon: Experiments with interferon in man. Lancet **1**, 505—506 (1965).

SHIBLEY, G. S., F. M. HANGER, and A. R. DOCHEZ: Studies in the common cold. 1. Observations of the normal bacterial flora of nose and throat with variations occurring during colds. J. exp. Med. **43**, 415—431 (1926).

SPIGLAND, I., J. P. FOX, L. R. ELVEBACK, F. E. WASSERMAN, A. KETLER, C. D. BRANDT, and A. KOGON: The virus watch program: a continuing surveillance of viral infection in metropolitan New York families. II. Laboratory methods and preliminary report on infections revealed by virus isolation. Amer. J. Epidem. **83**, 413—435 (1966).

STENHOUSE, A. C.: Rhinovirus infection in acute exacerbations of chronic bronchitis: a controlled prospective study. Brit. med. J. **3**, 461—463 (1967).

STOTT, E. J.: (1967). In press.

STOTT, E. J., and D. A. J. TYRRELL: Some improved techniques for the study of rhinoviruses using HeLa cells. Arch. ges. Virusforsch. **23**, 236—244 (1968).

STOTT, E. J., and M. WALKER: Human embryo kidney fibroblasts for the isolation and growth of rhinoviruses. Brit. J. exp. Path. **48**, 544—551 (1967).

Sutton, R. N. P., and D. A. J. Tyrrell: Some observations on interferon prepared in tissue cultures. Brit. J. exp. Path. 42, 99—105 (1961).

Taylor-Robinson, D.: Studies on some viruses (rhinoviruses) isolated from common colds. Arch. ges. Virusforsch. 13, 281—293 (1963a).

Taylor-Robinson, D.: Laboratory and volunteer studies on some viruses isolated from common colds (rhinoviruses). Amer. Rev. resp. Dis. 88, 262—268 (1963b).

Taylor-Robinson, D.: Respiratory virus antibodies in human sera from different regions of the world. Bull. Wld Hlth Org. 39, 833—847 (1965).

Taylor-Robinson, D., and M. L. Bynoe: Inoculation of volunteers with H rhinoviruses. Brit. med. J. 1, 540—544 (1964).

Taylor-Robinson, D., and N. Dimmock: Unpublished 1964, quoted in Tyrrell (1965).

Taylor-Robinson, D., R. Hucker, and D. A. J. Tyrrell: Studies on the pathogenicity of tissue cultures of some viruses isolated from common colds. Brit. J. exp. Path. 43, 189—193 (1963).

Taylor-Robinson, D., K. M. Johnson, H. H. Bloom, R. H. Parrott, M. A. Mufson, and R. M. Chanock: Rhinovirus neutralizing antibody responses and their measurement. Amer. J. Hyg. 78, 285—292 (1963).

Taylor-Robinson, D., and D. A. J. Tyrrell: Serotypes of viruses (rhinoviruses) isolated from common colds. Lancet 1, 452—454 (1962a).

Taylor-Robinson, D., and D. A. J. Tyrrell: Serological studies on some viruses isolated from common colds (rhinoviruses). Brit. J. exp. Path. 43, 264—275 (1962b).

Taylor-Robinson, D., and D. A. J. Tyrrell: Virus diseases on Tristan da Cunha. Trans. roy. Soc. trop. Med. Hyg. 57, 19—22 (1963).

Thomson, D., and R. Thomson: The common cold. Annals Pickett Thomson Research Laboratory 8, 1—738 (1932).

Tyrrell, D. A. J.: Common cold viruses. Int. Rev. exp. Path. 1, 209—242 (1962).

Tyrrell, D. A. J., in: Perspectives in Virology, vol. 3, edited M. Pollard, Hoeber, New York (1963a).

Tyrrell, D. A. J.: The use of volunteers. Amer. Rev. resp. Dis. 88, 128—134 (1963b).

Tyrrell, D. A. J.: The common cold and related diseases. Ed. Arnold London (1965).

Tyrrell, D. A. J., and C. J. Blamire: Improvements in a method of growing respiratory viruses in organ cultures. Brit. J. exp. Path. 48, 217—227 (1967).

Tyrrell, D. A. J., and M. L. Bynoe: Inoculation of volunteers with JH strain of a new respiratory virus. Lancet 2, 931—933 (1958).

Tyrrell, D. A. J., and M. L. Bynoe: Some further virus isolations from common colds. Brit. med. J. 1, 393—397 (1961).

Tyrrell, D. A. J., and M. L. Bynoe: Cultivation of a novel type of common cold virus in organ cultures. Brit. med. J. 1, 1467—1470 (1965).

Tyrrell, D. A. J., M. L. Bynoe, F. E. Buckland, and L. Hayflick: The cultivation in human-embryo cells of a virus (DC) causing colds in man. Lancet 2, 320—322 (1962).

Tyrrell, D. A. J., M. L. Bynoe, G. Hitchcock, H. G. Pereira, and C. H. Andrewes: Some virus isolations from common colds. 1. Experiments employing human volunteers. Lancet 1, 235—237 (1960).

Tyrrell, D. A. J., and R. M. Chanock: Rhinoviruses: a description. Science 141, 152—153 (1963).

Tyrrell, D. A. J., S. K. R. Clarke, R. B. Heath, R. C. Curran, T. S. L. Beswick, and L. Wolman: Studies of a Coxsackie virus antigenically related to ECHO 9 virus and associated with an epidemic of aseptic meningitis with exanthem. Brit. J. exp. Path. 39, 178—191 (1958).

Tyrrell, D. A. J., B. Head, and M. Dimic: Titration of tuberculin, nucleic acids and viruses by haemaggregation inhibition using a "pattern" test. Brit. J. exp. Path. 48, 513—621 (1967).

TYRRELL, D. A. J., and R. PARSONS: Some virus isolations from common colds. III. Cytopathic effects in tissue cultures. Lancet 1, 239—242 (1960).

TYRRELL, D. A. J., and B. RIDGEWELL: Freeze-drying of certain viruses. Nature (Lond.) 206, 115—116 (1965).

Virus Subcommittee of International Nomenclature Committee: Recommendations on virus nomenclature. Virology 21, 516—517 (1963).

WALLIS, C., and J. L. MELNICK: Cationic stabilization — a new property of enteroviruses. Virology 16, 504—505 (1962).

WEBB, P. A., K. M. JOHNSON, and M. A. MUFSON: A description of two newly-recognized rhinoviruses of human origin. Proc. Soc. exp. Biol. (N.Y.) 116, 845—852 (1964).

WHEELOCK, E. F., and W. A. SIBLEY: Interferon in human serum during clinical viral infections. Lancet 2, 382—385 (1964).

WIZIGMANN, G., und B. SCHIEFER: Isolierung von Rhinoviren bei Kälbern und Untersuchungen über die Bedeutung dieser Viren für die Entstehung von Kälbererkrankungen. Zbl. Vet.-Med. 13, 37—50 (1966).

Working Party: A collaborative study of the aetiology of acute respiratory infections in Britain 1961—1964. A report of the Medical Research Council working party on acute respiratory virus infections. Brit. med. J. 2, 319—326 (1965).

ZWILLENBERG, L. O., and F. BÜRKI: On the capsid structure of some small feline and bovine RNA viruses. Arch. ges. Virusforsch. 19, 373—384 (1966).